Wildfire and Water Quality: Processes, Impacts and Challenges

Recent IAHS Publications

Revisiting Experimental Catchment Studies in Forest Hydrology

Editors A. A. Webb, M. Bonell, L. Bren, P. N. J. Lane, D. McGuire, D. G. Neary, J. Nettles, D. F. Scott, J. D. Stednick & Y. Wang

Publ. 353 (April 2012) ISBN 978-1-907161-31-5, 240 + viii pp. £56.00

Most of what we know about the hydrological role of forests is based on paired catchment experiments whereby two neighbouring forested catchments are jointly monitored during a calibration period of several years, after which one of the catchments is kept untouched as a reference (control), while the second is submitted to a forest treatment (impact). This volume, generated from a workshop that gathered forest hydrologists from around the world, with the aim of revisiting results and promoting a renewal of international collaboration on this topic, is divided into four sections:

1 Addressing new questions using historical data sets
2 Impacts of fires
3 Water quality and sediment loads
4 Ecosystem services

Sediment Problems and Sediment Management in Asian River Basins

Editor Des E. Walling

Publ. 349 (2011) ISBN 978-1-907161-24-7, 224 + viii pp. Price £52.00

Sediment problems are assuming increasing importance in many Asian river basins and can represent a key impediment to sustainable development. Such problems include accelerated soil erosion, reservoir sedimentation and the wider impact of sediment on aquatic ecology, river morphology and water resource exploitation. They are further complicated by the impact of climate change and other components of global change in causing both increases and decreases in the sediment load of many rivers in recent years. In order to address these problems, sediment management must be seen as a central component of integrated river basin management. This volume, arising from a workshop organised by the International Commission on Continental Erosion (ICCE) of IAHS, the UNESCO International Sediment Initiative (ISI) and the World Association for Sedimentation and Erosion Research (WASER), focuses on sediment problems in Asian river basins and the many difficulties involved in their effective management. The first section comprises overviews of the sediment problems experienced by individual countries or particular issues relating to the wider region; and the second documents case studies that deal with specific problems and their management.

Water Quality: Current Trends and Expected Climate Change Impacts

Editors Norman E. Peters, Valentina Krysanova, Ahti Lepistö, Rajendra Prasad, Martin Thoms, Rob Wilby & Sarantuyaa Zandaryaa

Pub. 348 (June 2011) ISBN 978-1-907161-23-0, 186 + x pp. £50

The outcome of a symposium which brought together water quality scientists for a dialogue on the evaluation of climate change impacts on a wide range of water quality issues: Seasonality and extreme event effects on water quality; Effects on groundwater quality; Climate change and water quality assessment; Climate change and water temperature; and Climate change and water quality modelling.

BENCHMARK PAPERS IN HYDROLOGY

The IAHS Series that collects together, by theme, the papers that provided the scientific foundations for hydrolgy in stand-alone volumes.

RIPARIAN ZONE HYDROLOGY AND GEOCHEMISTRY

T. P. Burt, G. Pinay & S. Sabater

BM5 ISBN 978-1-907161-09-4 (2010) A4 format, hardback, 490 pp, £65.00

Study specifically of riparian zones is relatively new in hydrology, and while the oldest of the 36 benchmark papers selected for this volume dates to 1936, others were published in the 1970s and 1980s. They are grouped under the topics: Landscape ecology, Hydrology of the riparian zone, Linking riparian zone hydrology to solute transport, Biogeochemical processes and methods, Riparian buffering of surface and subsurface flows, and In-stream processes. Together, the papers and the editors' commentaries map the breakthroughs in the development of this important subdiscipline.

HYDRO-GEOMORPHOLOGY, EROSION AND SEDIMENTATION

by *Michael J. Kirkby*

BM6 ISBN 978-1-907161-14-8 (2011) A4 format, hardback, 640 pp, £70.00

A systematic analysis of the relationships between hydrology and geomorphology with commentaries on the papers that have been most influential in the development of research at the hydrology/geomorphology interface. Thirty-seven papers are reprinted in full or in part, the majority published pre-1970, including early contributions by Fisher (1866), Davison (1889) and Gilbert (1909), and seminal papers by Hack, Strahler, Wolman & Miller, and Melton, among others.

FOREST HYDROLOGY

by *David R. DeWalle*

BM7 ISBN 978-1-907161-17-9 (2011) A4 format, hardback, 474 pp, £65.00

The papers selected include the early review of forest and water by Zon (1927) and the Wagon Wheel Gap paired watershed study (Bates & Henry, 1928) report, but covers all aspects from evapotranspiration to water yields and quality to soil erosion.

Sponsored by SAHRA, University of Arizona

IAHS Publications can be ordered from the online bookshop at
www.iahsmembers.info/shop.php

or by contacting:
IAHS Press, CEH Wallingford, Oxfordshire OX10 8BB, UK
email: jilly@iahs.demon.co.uk
tel: +44 (0) 1491 692442 fax: +44 (0) 1491 692448

Full details of publications available at www.iahs.info

Edited by:

MIKE STONE
University of Waterloo, Ontario, Canada

ADRIAN COLLINS
ADAS, Wolverhampton, UK

MARTIN THOMS
University of New England, Armidale, Australia

IAHS Publication 354
in the IAHS Series of Proceedings and Reports

Published by the International Association of Hydrological Sciences 2012

IAHS Publication 354

ISBN 978-1-907161-32-2

British Library Cataloguing-in-Publication Data.
A catalogue record for this book is available from the British Library.

The papers included in this volume have been reviewed and some were extensively revised by the Editors, in collaboration with the authors, prior to publication.

IAHS is indebted to the employers of the Editors for the invaluable support and services provided that enabled them to carry out their task effectively and efficiently.

Publications in the series of Proceedings and Reports are available from:
IAHS Press, Centre for Ecology and Hydrology, Wallingford, Oxfordshire OX10 8BB, UK
tel.: +44 1491 692442; fax: +44 1491 692448; e-mail: jilly@iahs.demon.co.uk

Printed by Berforts Information Press

Cover images All photographs by Dr Uldis Silins (University of Alberta, Canada).

Preface

There has been increasing global concern over the impacts of landscape disturbance by wildfire on a range of aquatic ecosystem services and drinking water supply. Profound and often irreversible changes in river ecosystem function, geomorphology, water quality and water supply occur due to the severity and magnitude of wildfire-related landscape disturbance. Such impacts have important management implications for source water supply and protection at the catchment scale.

A conference on *Wildfire and Water Quality: Processes, Impacts and Challenges* was held in Banff Alberta, Canada, 11–14 June 2012, to bring together researchers and practitioners from diverse fields of hydrology, sediment transport, water quality and watershed management. The goals of the symposium were to improve knowledge of the impacts of large-scale landscape disturbances by wildfire on freshwater ecosystems and better elucidate processes that influence the source, transport and fate of sediment-associated contaminants in the aquatic environment. The symposium was sponsored by the International Committee on Continental Erosion (ICCE) of the International Association for Hydrological Sciences (IAHS), the Government of Alberta, Alberta Innovates – Energy and Environment Solutions, ADAS (UK) and the University of Waterloo.

This IAHS publication contains a selection of peer-reviewed papers presented at the symposium and contains 15 papers from six countries, reflecting the international dimension of the symposium. The symposium addressed several key themes and provided an opportunity for state-of-the-art knowledge transfer and exchange regarding the impacts of large-scale landscape disturbance by wildfire on water quality and its implications for sustainable management of water-sediment systems. Specific themes addressed in this volume include: (1) impacts of wildfire on hillslope hydrology, (2) effects of wildfire on the physical, chemical and biological composition of soils, (3) changes in sediment transport dynamics and yields resulting from wildfires, (4) methodologies used to evaluate the provenance and fate of wildfire impacted sediments and associated contaminants, (5) prediction of hydrological and sediment transport recovery trajectories at the local and catchment scale, (6) impacts of wildfire on aquatic ecology, (7) post-fire sedimentation and water quality impacts in reservoirs, and (8) management actions to reduce the impact of wildfires or river ecosystems.

We gratefully acknowledge Cate Gardner and her staff at IAHS Press, Wallingford, for their support with this volume.

Editors

Mike Stone
Department of Geography and Environmental Management, University of Waterloo, Ontario, Canada

Adrian Collins
ADAS, Woodthorne, Woverhampton, UK

Martin Thoms
Riverine Landscapes Research Laboratory, University of New England, Armidale, Australia

Contents

The effects of wildfire on sediment-associated phosphorus forms in the Crowsnest River basin, Alberta, Canada

DON ALLIN[1], MICHEAL STONE[1], ULDIS SILINS[2], MONICA B. EMELKO[3] & ADRIAN L. COLLINS[4]

1 *Department. of Geography and Environmental Management, University of Waterloo, Waterloo, Ontario N2L 3G1, Canada*

2 *Department of Renewable Resources, University of Alberta, Edmonton, Alberta T6G 2H1, Canada*

3 *Department of Civil and Environmental Engineering, University of Waterloo, Waterloo, Ontario N2L 3G1, Canada*

4 *Soils, Crops and Water, ADAS, Woodthorne, Wergs Road, Wolverhampton WV6 8TQ, UK*

Abstract The impacts of large-scale land disturbance by wildfire on a wide range of water and related ecological services are increasingly being recognized worldwide. This study explores the long-term impact (6–7 years) of the 2003 Lost Creek wildfire on particulate phosphorus forms (NAIP, AP, OP) of suspended river sediment at a large regional scale (554 km^2) in the Crowsnest River basin, Alberta, Canada. While total P concentrations were similar among burned and unburned river sediments, the mean bioavailable NAIP fraction remained approximately 70% greater and the organic P over 2-fold higher in sediments from five burned tributary watersheds compared to the reference site in the Crowsnest River study catchment. Because of the key role of phosphorus in regulating aquatic productivity in oligotrophic mountain rivers, these findings highlight the risk of a large scale and long-term legacy of wildfire in some mountain river systems.

Key words wildfire; phosphorus speciation; bioavailability; propagation

INTRODUCTION

Phosphorus (P) enrichment of surface waters has resulted in the eutrophication of many freshwater environments (Wall *et al.*, 1982; Cooke *et al.*, 1986). This global water quality problem has had significant impacts on aquatic ecology (Carpenter *et al.*, 1998; Correll, 1998) and the cost of water treatment and supply (Emelko *et al.*, 2010). Elevated P levels in rivers can increase the growth of periphyton and epiphytic diatoms, which often accelerate biofilm growth (House, 2003). The incidence and severity of this environmental problem varies spatially and temporally, and is governed by a number of factors that control the source, redistribution and bioavailability of P (Withers & Jarvie, 2008). Diffuse agricultural and urban sources represent the predominant inputs of dissolved and particulate P (PP) loading to receiving freshwaters (Dillon & Kirchner, 1978; Hill, 1981; Bird, 1986; O'Driscoll *et al.*, 2010). Other factors such as precipitation (Macrae *et al.*, 2010) and geology (Grobler & Silberbauer, 1981) influence the source, delivery pathways (surface and subsurface), composition (concentration, speciation and bioavailability) and fate of P in aquatic systems (Withers & Jarvie, 2008).

Phosphorus production in most undisturbed forested landscapes is usually low (Burke *et al.*, 2005); however, disturbance of forests by wildfire can dramatically increase sediment erosion rates and sediment-associated nutrient fluxes from fire-affected forests. Smith *et al.* (2011) report that wildfire can increase P production in streams from 0.3 to over 5 times greater than unburned conditions. Furthermore, plot scale studies suggest the relative bioavailability of P may increase in burned soils (Blake *et al.*, 2009, 2010). While sediment is the primary vector for the delivery of P to receiving freshwaters along with subsequent transport and storage and remobilisation within streams and rivers, comparatively few studies have explored post-fire P dynamics at larger basin scales. Given growing concern about the potential downstream effects of wildfire on water quality in many regions, there is a need to study these coupled sediment–nutrient dynamics in larger wildfire-affected river systems. The source waters at risk of wildfire represent essential water supplies for many cities and their protection from enhanced nutrient inputs is therefore desirable.

Sequential extraction techniques have been widely used to determine phosphorus (P) forms in aquatic sediments (Logan *et al.*, 1979; Ostrofsky 1987; Nürnberg 1988; Stone & English, 1993; Stone, 2004; Pacini & Gachter, 2008) and to estimate the release potential of P to the water column (Rydin, 2000; Reitzel *et al.*, 2005). Particulate P fractions are "operationally" defined by the extracting agent and the extraction conditions used (Psenner *et al.*, 1988). Sequential extraction techniques have been used to evaluate spatial and temporal variation in PP forms in rivers draining urban and agricultural landscapes (Logan *et al.*, 1979; Stone & English, 1993; Pacini & Gachter, 1999). However, no studies have been conducted specifically to evaluate PP forms in rivers draining landscapes disturbed by wildfire. Given the global increase in wildfire (Westerling *et al.*, 2006) and its potential impacts on water supply and treatment (Emelko *et al.*, 2010), there is a need to characterize sediment-associated P forms in streams disturbed by wildfire in order to assess their bioavailability and potential impact on downstream water quality. Accordingly, the objective of this study is to quantify the effects of wildfire on the PP forms (NAIP, AP, OP) of suspended river sediment at a larger, regional basin scale (554 km^2) in southern Alberta, Canada. Some implications of the study for downstream water quality and its management are discussed.

METHODS

Study area

The study was conducted in the Crowsnest River basin in southwestern Alberta, Canada (Fig. 1). Elevations in the region range from 1100 to 3100 m and vegetation consists primarily of mixed conifer forests at lower elevations and subalpine forests at mid elevations. At higher elevations, alpine ecozones are characterized by alpine meadow vegetation and bare rock extending above the tree line. Annual precipitation varies from 700 to 1700 mm, approx. 55% of which falls as rain during the frost free period. The mean annual (2000–2009) river discharge in the Crowsnest River (gauged at Frank, AB between tributaries 3 and 4; Fig. 1) was 341 mm and the annual hydrograph reflects a strong snowmelt-dominated seasonal response along with significant groundwater contributions to baseflows (Rock & Mayer, 2006).

In 2003, the Lost Creek Fire generated a near contiguous crown fire that consumed nearly all forest cover over 21 000 ha in the headwater regions of the Castle and Crowsnest rivers. Unpublished data for 2004–2011 indicates that the concentration, export and yields of both sediment and P in the burned watersheds have yet to recover to pre-burn conditions, thus illustrating the severity and prolonged impact of this mass landscape disturbance on sediment and P fluxes. The Lost Creek wildfire occurred immediately upstream of the Oldman Dam (Fig. 1) and there is concern regarding the water quality implications of enhanced sediment and P fluxes to the reservoir from burned landscapes.

Sample collection

Composite samples of suspended solids were collected passively with a network of *in situ* time-integrating samplers (Phillips *et al.*, 2000). These samplers provide a simple and pragmatic means of sampling the natural variation in sediment properties during snowmelt and storm events in rivers (Phillips *et al.*, 2000) and are routinely used to collect a sufficiently large sample mass for laboratory analyses in studies designed to evaluate the geochemical and contaminant properties of fluvial sediment (e.g. Walling *et al.*, 2008).

Time-integrating samplers were deployed in the Crowsnest River at locations upstream and downstream of the wildfire and in five wildfire impacted streams tributary to the Crowsnest River (Fig. 1). The upstream sample location on the Crowsnest River (site 1) represents a relatively undisturbed (reference) sub-catchment. The tributary sampling locations (sites 2, 3, 4, 5, 6) drain catchments impacted to varying degrees (8 to 100%) by wildfire (Table 1). The downstream Crowsnest River location (site 7) was selected to examine the combined signatures of sediment P fluxes from the burned catchments on the sediment quality of the Crowsnest River at a large (554 km^2) basin scale. Time-integrating samplers were installed each spring prior to snowmelt and composite sediment samples were collected three times in 2009 and twice in 2010. Sediment samples were frozen for storage prior to chemical analysis.

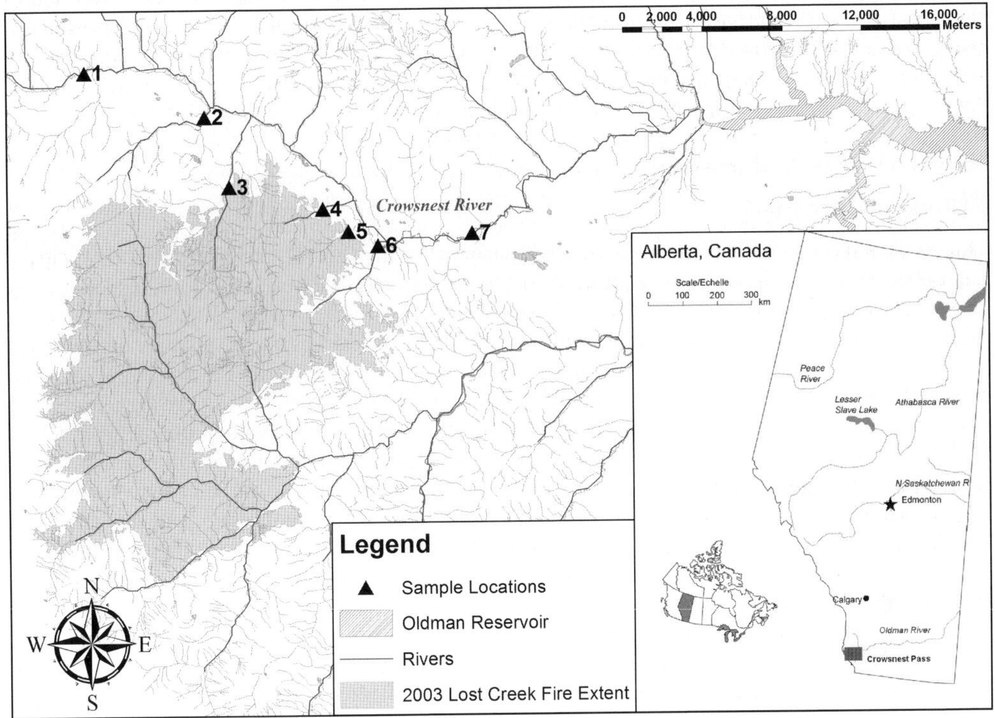

Fig 1 The study area and river suspended sediment sampling locations.

Table 1 Characteristics of the study sub-catchments.

Site	River	Upstream drainage area (ha)	Upstream burn Area (ha)	% Upstream burned	% Upstream salvaged logged
Reference					
1	Crowsnest River	16 076	0	0	-
Burned tributaries					
2	York Creek	3 365	271	8	-
3	Lyons Creek	2 650	1 703	64	21
4	Drum Creek	1 179	1 064	90	-
5	Unnamed Creek	478	450	94	-
6	Byron Creek	2 511	1 508	60	5
Downstream mainstem					
7	Crowsnest River	55 387	5 300	10	1

Particulate P fractionation

A sequential extraction scheme was used to quantify the relative fractional composition of PP forms in sediment samples. The fractionation scheme yields five operationally-defined fractions of PP (Stone & English, 1993). In order of extraction, these PP fractions are: (a) loosely sorbed P; (b) reductant soluble P; (c) reactive P sorbed to metal oxides; (d) P bound to carbonates, apatite P and P released by the dissolution of oxides; and (e) non-reactive organic P extractable in hot (85°C) NaOH. The first three fractions combined are considered to be bioavailable and referred to collectively as non-apatite inorganic P (NAIP). Fractions four and five are referred to as apatite P

(AP) and organic P (OP), respectively. The OP fraction is potentially available after mineralization (Bostrom *et al.*, 1988). After centrifugation, extracts were analysed on a Technicon Autoanalyzer using the stannous chloride ammonium molybdate method (Environment Canada, 1979). The detection limit of the analytical method is 1 μg/L.

RESULTS AND DISCUSSION

Phosphorus speciation

Due to the importance of sediment-associated transport for the flux of nutrients within river basins, a number of studies have been conducted to characterize the PP forms in both suspended and deposited sediment in rivers and lakes. Owens & Walling (2002) reported that the phosphorus content of fluvial sediment in rural and industrialized river basins typically ranges between 100 and 13 500 μg/g. In a variety of calcareous and non-calcareous lakes, total PP concentrations were reported to range from 580 to 7000 μg/g (Williams *et al.*, 1976; Peterson *et al.*, 1988; Ostrofsky & McGee, 1991; White & Stone, 1996). In the present study, the total PP content ranged from 543 to 787 μg/g (Fig. 2), which is much lower than the PP levels commonly measured in rivers and lakes directly impacted by urban and agricultural land use. However, the undisturbed headwater streams on the eastern slopes of the Rocky Mountains are typically oligotrophic and increased TP concentrations due to wildfire in the study area have dramatically increased stream productivity and biofilm growth (Silins *et al.*, 2009). The increased biofilm growth has decreased the erodibility of cohesive sediment deposits in wildfire affected streams (Stone *et al.*, 2010).

Longitudinally, the average concentration of the NAIP fraction in the Crowsnest River increased from 163 μg/g at the upstream (site 1) to 386 μg/g at the downstream sampling location (site 7; Fig. 2). As a percentage of total P, NAIP increased by a factor of 2 from 26% to 52.5% (Table 2). The NAIP fraction in the burned tributary sediments (sites 4, 5, 6) and the most downstream location on the Crowsnest River (site 7) was greater than the upstream reference site. With the exception of Lyons Creek, the NAIP content increased with the spatial extent of burn in the corresponding sub-catchment. Accordingly, the highest average NAIP fraction concentration was observed in Drum Creek, which experienced a 90% burn during the 2003 wildfire.

The bioavailability of sediment-associated P for plant uptake depends upon several interrelated physical, chemical and biological factors (Bird, 1986). Grain size is one factor that strongly influences the major element composition and distribution of PP forms in aquatic sediment (Stone & Mudroch, 1989; Stone & English, 1993). The median diameter (D_{50}) of suspended solids collected from the reference, burned and downstream Crowsnest River sample stations ranged from 16 to 59 μm. The NAIP fraction associated with these fine-grained materials indicated that the potential bioavailability of sediment-associated P in the tributary sediment from the burned catchments is 1.2 to 2.6 times higher than the upstream reference site on the Crowsnest River.

The average AP content in sediment was much greater in the upstream reference site than at either the burned tributaries or most downstream site on the Crowsnest River (Fig. 2). The respective AP fraction of PP decreased from 61% at the reference site to 38% at the downstream Crowsnest River site. A dilution of the AP signal at site 7 is related to lower AP content from tributary inputs, particularly Drum Creek, which has high sediment yields but a low AP fraction (26% of total P). DePinto *et al.* (1981) measured algal-available phosphorus in suspended sediments collected from the Maumee, Sandusky and Cattaraugus rivers. They observed that total P content varied by less than 5% from levels that were reported in an earlier investigation of the same rivers. Accordingly, the authors suggested that PP content may be a temporally stable characteristic for any given tributary. Similarly, Stone & English (1993) observed low seasonal variability in the AP fractions in sediment collected from two southern Ontario rivers. Their observation supports the contention of Thomas & Munawar (1985) that AP is a constant background form of P in the tributaries of the lower Great Lakes. Burrus *et al.* (1990) examined the seasonal delivery of PP forms to Lake Geneva from the upper Rhone River and found that AP content remained constant throughout the year. In the present study, the AP content was measured

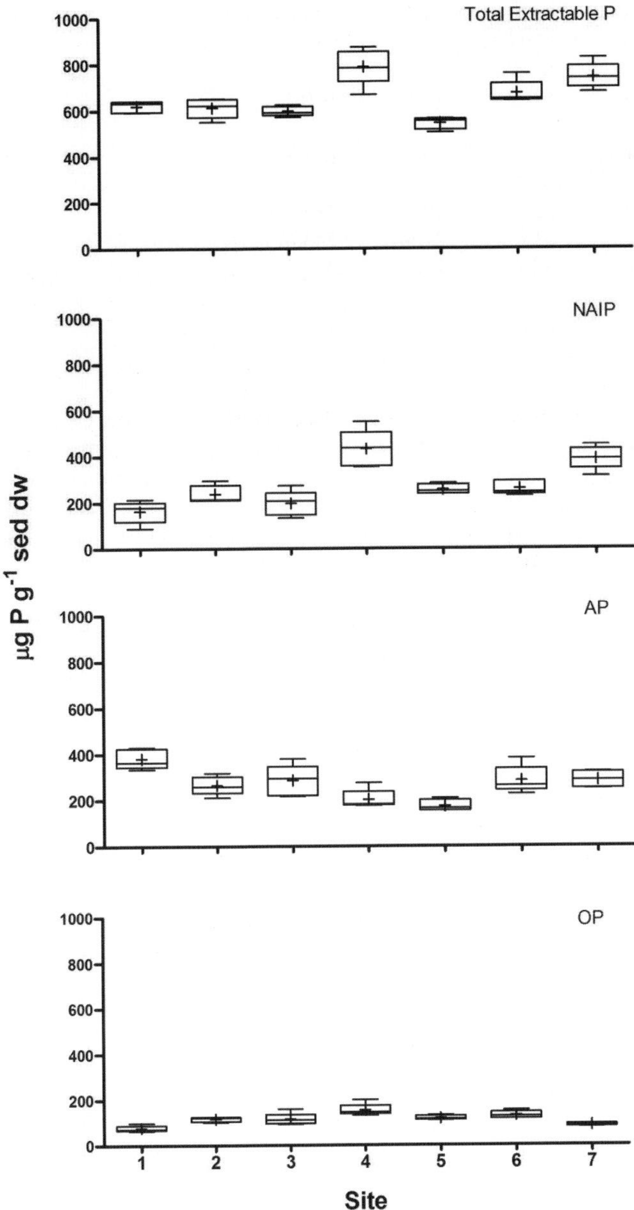

Fig. 2 Distribution of total extractable PP, non-apatite inorganic P, apatite inorganic P, and organic P among sample sites across the 2009 and 2010 collection periods. Horizontal line indicates median, boxes indicate 25th and 75th percentiles, whiskers indicate range, and + indicates the mean.

seven times over a two year period, but varied by <0.5%. The high AP concentration observed at site 1 indicates there is a strong geological control on the PP fractionation in the catchment above the reference sediment sampling site. Given the low variability of AP at site 1, it can be expected that any variation in total PP export will likely reflect variation in the NAIP and OP fractions resulting from land use disturbance in the Crowsnest River such as wildfire and salvage logging.

Table 2 NAIP, AP and OP as a percentage of total PP for the study sites.

Site	*n*	% NAIP of TP	% AP of TP	% OP of TP	TP
Reference					
1	7	26.3	61.3	12.5	619.2
Burned tributaries					
2	7	38.8	43.2	19	611.5
3	6	32.9	47.7	19.4	596.3
4	5	54.8	25.7	19.6	787
5	6	46.8	31.8	21.8	543.6
6	6	38.5	42.1	19.4	671.7
Downstream mainstem					
7	9	52.1	38.4	11.9	741.2

The average OP fraction in the Crowsnest River increased slightly from site 1 (77 μg/g) to (88 μg/g) at site 7 (Fig. 2). However, there was a 1.5- to 2-fold increase in the OP fraction of the tributary sediment compared to the upstream reference site. Compared to streams draining unburned landscapes, the impacted tributaries have experienced accelerated and sustained algal/biofilm growth since the Lost Creek wildfire in 2003 (Silins *et al.*, 2009). The higher OP particulate fraction observed in the burned tributaries likely includes organic materials eroded and remobilized from the stream bed during higher flow events.

While the impact of wildfire on nutrient production has received increasing research attention over the past decade, the issue of potential impacts on larger river systems downstream of wildfires remains uncertain. Hauer & Spencer (1998), Burke *et al.* (2005), and Mast & Clow (2008) all reported significant impacts of wildfires on larger river systems (106–248 km^2) that were detectable up to 4 years after wildfire. In the present study, while the 2003 Lost Creek wildfire burned a large region of the headwater between the Crowsnest and Castle river basins, the fire affected region of the Crowsnest River at site 7 only comprised 10% of a 554 km^2 basin. Moreover, the signature of this wildfire on increased bioavailable P forms was clearly evident in suspended river sediments 6 and 7 years after the wildfire of 2003. The foregoing supports the idea that sediment-associated contaminant storage and transport downstream of wildfire affected regions is an important component governing the spatial scale and longevity of potential downstream effects of wildfires that will need increased attention to understand better the range of impacts of wildfires on lotic ecosystems. The observed legacy of elevated sediment and associated NAIP delivery from burned landscapes to the Crowsnest River and downstream aquatic environments will have implications on water quality in the Oldman Reservoir. The eastern slopes of the Rocky Mountains are critical source water regions in Alberta. The potential impact of wildfire on a range of ecosystem services and long-term water supply to municipalities, associated with increased sediment-associated nutrient pressures, is a potential concern requiring well-integrated landscape-level planning and management programmes.

CONCLUSIONS

The results of this study clearly showed a longitudinal increase in the bioavailable NAIP fraction of suspended solids downstream of a burned region of the Crowsnest River basin, relative to a reference site upstream of the 2003 Lost Creek wildfire. The storage and transport of NAIP-rich sediments is likely implicated in the longevity of the effects we observed 6–7 years after the 2003 wildfire and, in turn, these processes may have important implications for water quality in the Oldman Reservoir further downstream. The longer-term propagation of landscape disturbance impacts resulting from wildfire requires further research to continue reinforcing findings such as those reported here.

Acknowledgements The authors gratefully acknowledge the field support of C. Williams, M. Wagner, A. Martens, A. Peter-Rennich, J. Fitzpatrick and E. Noton. Funding for the research was provided by AWRI/AIEES, and NSERC grants to M. Stone. Support for this research by Alberta Sustainable Resource Development (forest management division) is also gratefully acknowledged.

REFERENCES

Bird, G. A. (1986) Phosphorus dynamics in Great Lakes ecosystems. Environment Canada. 161 pp.

Blake, W. H., Wallbrink, P. J. & Droppo, I. G. (2009) Sediment aggregation and water quality in wildfire-affected river basins. *Marine & Freshwater Research* 60, 653–659.

Blake, W. H., Theocharopoulos, S. P., Skoulikidis, N., Clark, P., Tountas, P., Hartley, R. & Amaxidis, Y. (2010) Wildfire impacts on hillslope sediment and phosphorus yields. *J. Soils Sediments* 10, 671–682.

Burke, J. M., Prepas, E. E. & Pinder, S. (2005) Runoff and phosphorus export patterns in large forested watersheds on the western Canadian Boreal Plain before and for 4 years after wildfire *J. Environ. Eng. Sci.* 4, 319–325.

Bostrom, B., Persson, G. & Brombcre, B. (1988) Bioavailability of different phosphorus forms in freshwater systems. *Hydrobiologia* 170, 133–135.

Carpenter, S. R., Caraco, N. F, Correll, D. L., Howarth, R. W., Sharpley A. N. & Smith V. H. (1980) Nonpoint pollution of surface waters with phosphorus and nitrogen. *Ecological Applications* 8, 559–568.

Cooke, G. D., Welch, E. B., Peterson, S. A. & Newroth, P.R. (1986) *Lake and Reservoir Restoration*. Butterworths, Boston, USA.

Correll, D. L. (1998) The role of phosphorus in the eutrophication of receiving waters: A review. *J. Environ. Quality* 27(2), 261–266.

DePinto, J. V., Young, T. C. & Martin, S. C. (1981) Algal available phosphorus in suspended sediments from lower Great Lakes tributaries. *J. Great Lakes Res.* 7, 311–325.

Dillon, P. J. & Kirchner, W. B. (1975) The effects of geology and land use on the export of phosphorus from watersheds. *Water Research* 9, 135–148.

Environment Canada (1979) *Analytical Methods Manual*. Ottawa, Canada.

Fox, L. (1993) The chemistry of aquatic phosphate: inorganic processes in rivers. *Hydrobiologia ISi,* 1–16.

Grobler, D. C. &. Silberbauer, M. J. (1981) The combined effect of geology: phosphate sources and runoff on phosphate export from drainage basins *Water Research* 19(8), 975–981.

Hauer, F. R. & Spencer, C. N. (1998) Phosphorus and nitrogen dynamics in streams associated with wildfire: A study of immediate and longterm effects. *J. Wildland Fire* 8, 183–198.

Hill, A. R. (1981) Stream phosphorus exports from watersheds with contrasting land uses in southern Ontario. *Water Resources Bulletin* 17 (4), 627–634.

House, W. A. (2003) Geochemical cycling of P in rivers. *Appl. Geochem.* 18, 739–748.

Macrae M. L, English, M. C., Schiff, S. L. & Stone M. (2010) Influence of antecedent hydrologic conditions on patterns of hydrochemical export from a first-order agricultural watershed in Southern Ontario, Canada. *J. Hydrol.* 389, 101–110.

Mast, M. A. & Clow, D. W. (2008) Effects of 2003 wildfires on stream chemistry in Glacier National Park, Montana. *Hydrol. Processes* 22, 5013–5023.

Nürnberg, G. K. (1988) Prediction of phosphorus release rates from total and reductant- soluble phosphorus in anoxic lake sediments. *Can. J. Fish. Aquat. Sci.* 45, 453–462.

O'Driscoll, M., Clinton, S., Jefferson, A., Manda, A. & McMillan, S. (2010) Urbanization effects on watershed hydrology and in-stream processes in the southern United States. *Water* 2, 605–648;

Ostrofsky, M. L. (1987) Phosphorus species in the surficial sediments of lakes of Eastern North America. *Can. J. Fish. Aquat. Sci.* 44, 960–966.

Ostrofsky, M. & McGee, G. (1991) Spatial variation in the distribution of phosphorus species of in the surficial sediments of Canadohla Lake, Pennsylvania: Implications for internal phosphorus loading estimates. *Can. J. Fish. Aquat. Sci.* 48, 233–237.

Owens, P. N. & Walling, D. E. (2002) The phosphorus content of fluvial sediment in rural and industrialised river basins. *Water Research* 36, 685–701.

Peterson, K., Bostrom, B. & Jacobsen, 0. (1988) Phosphorus in sediment speciation and analysis. *Hydrobiologia* 170, 91–101.

Pettersson, K. & Istvanovics, V. (1988). Sediment phosphorus in Lake Balaton – forms and mobility. *Arch. Hydrobiol. Beih. Ergebn. Limnol.*, 30, 25–41.

Phillips, J. M., Russell, M. A. & Walling, D. E. (2000) Time-integrated sampling of fluvial suspended sediment: a simple methodology for small catchments. *Hydrol. Processes* 14, 589–602.

Reitzel, K., J., Hansen, F., Ø. Andersen, K. Hansen & Jensen, J. S. (2005) Lake restoration by dosing aluminum relative to mobile phosphorus in the sediment. *Environ. Sci. Technol.*39, 4134–4140.

Rock, L. & Mayer, B. (2006) Isotope hydrology of the Oldman River basin, southern Alberta, Canada. *Hydrol. Processes* 21, 3301–3315.

Rydin, E. (2000) Potentially mobile phosphorus in Lake Erken sediment. *Water Res*. 34, 2037–2042.

Silins, U., Bladon, K., Stone, M., Emelko, M., Boon, S., Williams, C., Wagner, M., & Howery, J. (2009) Southern Rockies Watershed Project: Impact of natural disturbance by wildfire on hydrology, water quality, and aquatic ecology of Rocky Mountain watersheds – Phase I (2004–2008), Report to Government of Alberta, 90 p.

Spiers, G., Dudas, M. & Turchenek, L. (1989) The chemical and mineralogical composition of soil parent material in northern Alberta. *Can. J. Soil Sci.* 69, 721–737.

Stone, M. (2004) Spatial distribution of particulate phosphorus forms in the Slave River Delta, Northwest Territories, Canada. In: *Sediment Transfer Through the Fluvial System* (ed. by V. Golosov *et al.*), 481–487. IAHS Publ. 288. IAHS Press, Wallingford, UK.

Stone, M. & Mudroch, A. (1989) The effect of particle size, chemistry and mineralogy of river sediments on phosphate adsorption. *Environ. Technol. Lett.* 10, 501–510.

Stone, M. & English, M. (1993) Geochemical composition, phosphorus speciation and mass transport characteristics of fine-grained sediment in two Lake Erie tributaries. *Hydrobiologia* 253, 17–29.

Stone, M. Emelko, M. B., Droppo, I. G. &Silins, U. (2010) Biostabilization and erodibility of cohesive sediment deposits in wildfire-affected streams. *Water Research* 45(2), 521–534.

Thomas, R.L & Munawar, M. (1985). The delivery and bioavailability of particulate bound phosphorus in Canadian rivers tributary to the Great Lakes. In: Proceedings of the International Conference Management Strategies for Phosphorus in the Environment (ed. by J. N. Lester & P. W. W. Kirk), 462–469.

Wall, G., Dickinson, T. & Van Vliet, L. (1982) Agriculture and water quality in the Canadian Great Lakes Basin: II Fluvial sediments. *Environ. Qual.* 11, 482–486.

Walling, D. E., Collins, A. L. & Stroud, R. (2008) Tracing suspended sediment and particulate phosphorus sources in catchments. *J. Hydrol.* 350, 274–289.

Wasson, R. J., Croke, B. F., McCulloch, M. M., Mueller, N., Olley, J., Starr, B., Wade, A., White, I. & Whiteway, T. (2003) Sediment, Particulate and Dissolved Organic Carbon, Iron and Manganese Input to Corin Reservoir. Report to ActewAGL for the Cotter Catchment Fire Remediation Project, WF 30014, Centre for Resource and Environmental Studies, Australian National University.

Westerling, A. L., Hildago, H. G., Cayan, D. R. & Swetnam, T. W. (2006) Warming and earlier spring increase Western US forest wildfire activity. *Science* 18(313), 940 943.

White, A. & Stone, M. (1996) Spatial variation and distribution of phosphorus forms in surficial sediments of two Canadian Shield lakes. *Can. Geogr.* 40(3), 258–265.

Williams, J., Murphy, T. & Mayer, T. (1976) Rates and accumulation of phosphorus forms in Lake Erie sediments. *J. Fish. Res. Bd. Can.* 33, 430–139.

Withers, P. J. & Jarvie, H. P. (2008) Delivery and cycling of phosphorus in rivers: A review. *Sci. Total Environ.* 400, 379–395.

The applicability of black carbon for tracing soil erosion: fire impacts on landscape dynamics in Cyprus

JENS BRAUNECK[1] & MANFRED LANGE[2]

1 *Physical Geography, University of Wuerzburg, Am Hubland, 97074 Wurzburg, Germany*
jens.brauneck@uni-wuerzburg.de

2 *Energy, Environment and Water Research Centre, Cyprus Institute, Nicosia, Cyprus*

Abstract On the Mediterranean island of Cyprus, a series of both natural and anthropogenic factors has led to severe land degradation in the past, which now results in water shortage during the summer months. A combined approach of terrain mapping, unmanned aerial system (UAS) flight missions and chemical characterization of black carbon will permit the classification of hazardous locations in terms of potential fires and erosion processes. The emphasis of the upcoming surveys is concentrated on those catchments that are connected to one of the numerous valley dammed reservoirs. There, the use of black carbon as an erosion indicator will be examined to trace the paths of eroded material inside the watershed and to enable a reconstruction of process dynamics and local fire history.

Key words Mediterranean; post-fire erosion; unmanned aerial system; black carbon; Cyprus

INTRODUCTION

According to the current scientific consensus, global climatic conditions will change considerably in the coming decades and centuries (IPCC, 2007). Global and regional climate models project significant temperature increases and reduced annual precipitation for the eastern Mediterranean, primarily during spring to summer. Consequently, this area is recognized as a hot spot for climatic change (Giorgi, 2006), where increasing annual mean temperatures will exacerbate problems with regard to water supply (Fox *et al.*, 2007) and the intensity of forest fires (Pausas, 1999). The risk of climatically-induced changes in environmental conditions across the Mediterranean region increases the need for low-cost, technically uncomplicated and versatile, practical methods for environmental monitoring that can deliver reliable spatial and temporal comparisons at national and international scales (Morvan *et al.*, 2008).

Forest fires and post-fire erosion have serious impacts on soil characteristics and are considered the main drivers of degradation of forested areas across the Mediterranean region. Although a natural feature of Mediterranean ecosystems (Naveh, 1975), fires have developed an increasingly destructive potential in recent decades, particularly in combination with unadapted land use (Naveh, 2007). In the 1990s, 600 000 hectares of forest were destroyed annually across the Mediterranean by forest fires (Shakesby & Doerr, 2006). The summer droughts responsible for increasing the risk of forest fires recently caused crop shortfalls in wide areas of the Mediterranean, for example in the Iberian Peninsula in 2005 (García Herrera *et al.*, 2007), and led to saving measures and emergency water supply such as in Cyprus in 2008 (Michaelides & Pashiardis 2008). For the Eastern Mediterranean, detailed studies on forest fires and post-fire erosion are rarely found, and those that do exist focus on a small number of countries (e.g. Israel, Greece). Area-wide and publicly-available analyses for Cyprus have not existed until now. In addition, most investigations in the region have not been published in peer-review journals, since they are published by local forestry and environmental authorities in reports for the purpose of documentation. The analysis of forest fires with regard to their impact on landscape dynamics can both contribute to the understanding of ecosystems and their morphological developments in the past, and to understanding their potential development under environmental perturbations in the future (Gedye *et al.*, 2000).

THE SITUATION IN CYPRUS

Cyprus (Fig. 1) is an island situated in the Eastern Mediterranean Sea, west of Syria, and is divided into the northern Turkish-controlled area and the larger Republic of Cyprus in the south. The

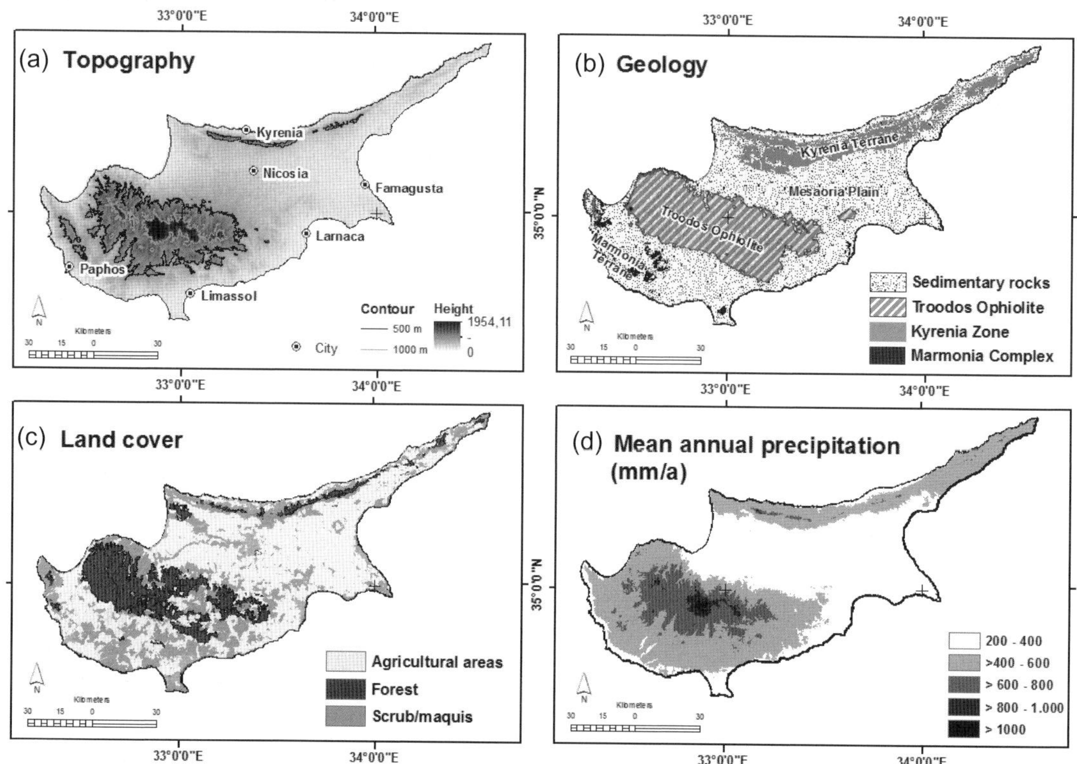

Fig. 1 Topography, geology, land cover and annual precipitation of Cyprus (Data sources: topography data from ASTER GDEM V2; land cover data from CLC2006; geology and precipitation data provided by Armin Duenkeloh/University of Wuerzburg).

geology of Cyprus (Fig. 1(b)) is dominated by the Troodos ophiolite, a sequence of former oceanic crust material, that now towers over the island up to 1952 m (Mount Olympus). The stratigraphy contains mantle and lava sequences as well as plutonic and sheeted dyke complexes (Varga & Moores, 1985). The much smaller Kyrenia range in the north consists of Permian to Miocene sedimentary rocks. A wide sedimentary plain is situated between these mountain ranges, and called the Mesaoria. The distribution and assemblage of soils indicates a long history of denudation and degradation. The Troodos mountain range is dominated by eutric-lithic Leptosols and eutric-skeletic Regosols. Eutric Cambisols and anthropic Regosols are primarily located in the valleys. The Mesaoria is dominated by calcic Luvisols, calcaric-lithic Leptosols and calcaric-leptic Regosols. Solitary occurrences of gleyic Solonets and vertic-leptic Calcisols are distributed all over the sedimentary rocks (Hadjiparaskevas, 2005).

Due to the Mediterranean climate, rainfall occurs from October to May. Annual precipitation is highly correlated with the topography of the island (Fig. 1(a) and (d)) but is also extremely variable (Fig. 2). Therefore, annual precipitation exceeds 1000 mm at some points in the Troodos Mountains, while the major part of the island can be considered semi-arid with values around 200–400 mm/year. The analysis of long-term data trends as well as the projections of regional climate models (RCM) both suggest an intensification of aridity and an increase of weather extremes across the eastern Mediterranean (Hadjinicolaou *et al.*, 2010).

In order to secure the water supply for domestic use and irrigation, more than 100 dams and reservoirs of different types, with a cumulative storage capacity of 300×10^6 m^3 were constructed in Cyprus during the last 70 years. The largest structure is the Kouris Dam with a capacity of

115×10^6 m^3. However, in 2008, after a series of four consecutive years with low precipitation, the storage volume of the major dams was reduced to less than 10%, meaning that the dams "were virtually empty" (Michaelides & Pashiardis, 2008). Due to these conditions, an expensive water transfer by boat from Greece had to be arranged to secure the domestic water supply and compensation payments were paid to farmers.

In the forested areas of the Troodos mountain range, wildfires and post-fire erosion exacerbate this critical situation by reducing the groundwater recharge rate of the aquifers located in this region. The most important aquifers are located in the fractured rocks of the ophiolite, as well as in sedimentary structures around the central Troodos. Re-occurring long-term droughts, a growing demand for water, as well as the reduced recharge of coastal aquifers due to the construction of dams has led to a reduction of the coastal groundwater tables and thus to an increased risk of sea water intrusion (European Environment Agency, 2009). Overall, 12 of 19 aquifers in southern Cyprus are already affected by, or are at risk of, saltwater intrusion. Severe salt water contamination has already occurred in the region of Larnaca, caused by the overexploitation of the Kiti aquifer for the purpose of agricultural irrigation (Milnes & Renard, 2004). In the southern part of the Troodos Mountains, the aquifers partially lie in the range of sedimentary structures; the groundwater there is already moderately saline. In the upper sections of the Kouris catchment, located in the central parts of the Troodos, depletion of groundwater due to irrigation methods has been reported (e.g. Boronina *et al.*, 2003).

About 19% of the island is covered with forest, but only half of that area is classified as high forest. State forests make up the major part (92% or 161 000 ha) and primarily consist of *Pinus brutia* (45%), reforestation areas (21%) and maquis vegetation (16.5%). Available fire statistics include all incidents in state forests. According to the forest fire statistics for 2001–2010, the main cause for forest fires on Cyprus was human activity, whether agricultural work (24%), fires caused by picnickers or travellers (16%), or arson (1%). Only 12% of all fires on Cyprus can be connected to natural causes such as lightning and 13% of all fires have unknown sources (Boustras *et al.*, 2008; Department of Forests, 2011). From 2000 to 2009, only 7% of all fires exceeded 5 ha in extent, and 83% affected only 1 ha, or less. The fires mainly occur in the fire season from April to November. Figure 2 shows the fire occurrence on Cyprus after a series of drought years. It can be argued that the large fire in 1998, as well as the increase of fire numbers in the subsequent years, is

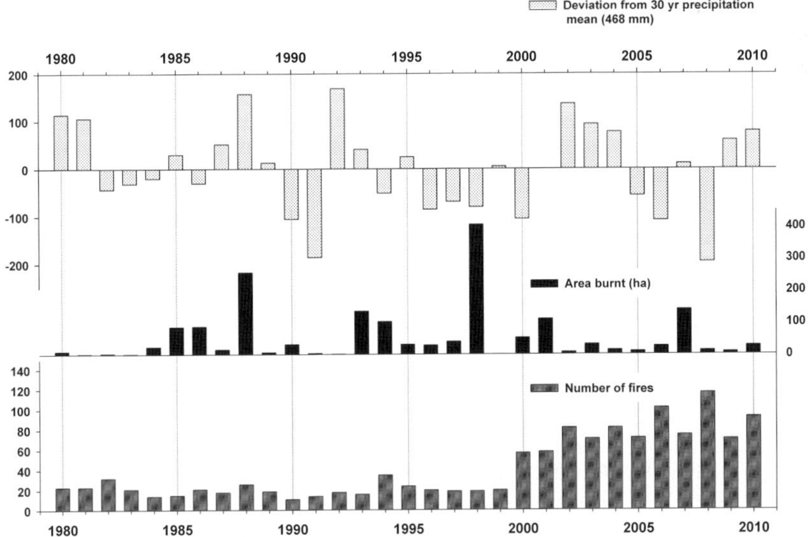

Fig. 2 Mean annual precipitation and fire statistics for Cyprus (1980–2010) (Data sources: Department of Forests, 2011; Cyprus Meteorological Service, 2011).

related to the increase in available fuel due to degraded vegetation. The obvious decrease in burnt area during the period 2001–2010, may be connected to the improvement and expansion of forest fire mitigation measures. These abatement options include the maintenance of forested areas, the construction of firebreaks, and the implementation of information campaigns for minimizing risky practices (Department of Forests, 2011).

POST-FIRE EROSION ASSESSMENT AND LOCAL FIRE RECONSTRUCTION METHODS

One primary object of fire-related surveys is assessing the impact of such landscape disturbance on erosion dynamics. Understanding the linkages between changing land cover and subsequent intensification of landscape dynamics have been a primary research objective in the Mediterranean for decades. Accordingly, a large number of articles cover these topics (see summaries of Boardman, 2006; Bakker *et al.*, 2008; García-Ruiz, 2010). Comparative studies of various locations across the Mediterranean have confirmed strong linkages between soil erosion rates and different land uses due to specific site parameters, the most important being vegetation cover (Cosmas *et al.,* 1997).

For any project that intends to analyse landscape dynamics, the mapping and monitoring of vegetation cover is therefore a fundamental task. Appropriate ground-cover information on a large scale can be derived from vegetation indices such as the Normalized Difference Vegetation Index (NDVI) that is commonly based on LANDSAT or NOAA data. But, due to the low resolution (60 m × 60 m) these data sources are not sufficient for applications at the erosion plot scale. Unmanned aerial systems (UAS) have the potential to link soil surveys and remote sensing data, especially in terms of "the scale and resolution gap" (Aber *et al.*, 2010). The application of unmanned aircraft can range from military use as drones to deployment in civilian projects, and these devices have been increasingly used for scientific purposes in recent years. Different kinds of UAS platforms can be equipped with a variety of sensors and are therefore adaptable to investigating many different scientific objectives. The most recent software developments even allow the generation of high-resolution elevation data from digital images (e.g. Haala *et al.*, 2010). Although numerous recent studies deal with the general issue of soil erosion across the Mediterranean region, the direct comparability of their results remains problematic due to the contrasting methods used to assemble the data. These methods include artificial irrigation and infiltration measurements (e.g. Seeger, 2007; Ries, 2010), permanently installed sediment traps and studies on retaining dams (e.g. Sirvent *et al.*, 1997; Romero Díaz *et al.*, 2007; Bellin *et al.*, 2009), and the conventional analysis of soil parameters on defined areas of investigation (plots) (e.g. Cammeraat, 2004; Brown *et al.*, 2009). Artificial tracer tests have, to date, been deployed less frequently due to the challenges associated with their implementation. For example, a recent study of soil erosion with synthetic magnetic tracers (Ventura *et al.*, 2002) reported the need to mimic the influence of size-dependent sorting during erosion processes. In this context, appropriate tracers require careful matching with the particle size characteristics of host soils and need to be able to mimic the natural transportability of soil particles taking account of density and particle interaction. As an alternative, natural environmental tracers seem to be more promising in terms of investigating fire signals. For example, the use of magnetism as a potential indicator of fire in sediment archives was investigated by Gedye *et al.* (2000). However, while magnetic peaks in sediment archives can be used as indications of fire, these do not allow the disentangling of the specific locations releasing soil and sediment post-fire.

In order to reconstruct system connectivity and sediment flux on a larger scale, geochemical fingerprinting can be used to establish the origin of deposited material. Rare earth elements are commonly used to specify individual source areas and to associate those to fractions within a sediment archive (e.g. Kimoto *et al.*, 2006). Fallout radionuclides are applicable for tracing erosion over recent time periods. The applicability of the different radionuclides (e.g. ^{137}Cs, unsupported ^{210}Pb and ^{7}Be) is determined by their specific half-life (e.g. Mabit *et al.*, 2008).

The immediate impact of forest fires is detectable on the basis of physical changes in vegetation, as well as in the alteration of soil parameters and properties (Koegel-Knabner, 2000), including pH, infiltration capacity or the stability of aggregates (Certini, 2005; Fox *et al.*, 2007; Knicker, 2007). The environmental impacts of forest fires have been the subject of a series of studies (e.g. Shakesby & Doerr, 2006; Cerdà & Doerr, 2008; Conedera *et al.*, 2009) using a range of research methods and techniques. According to Shakesby & Doerr (2006), the establishment of a uniform classification scheme for the extent of forest fires and their impacts on the soil surface is of particular relevance with regard to assisting the direct comparability of diverse studies. Since it is infeasible to measure the surface temperature during a natural fire, parameters that are recorded afterwards should be the basis for the categorization of various fire events. As the organic constituents of Mediterranean soils are heavily affected by fires in terms of both quality and quantity (Knicker *et al.*, 2006), they can be used to reconstruct the extent as well as the impact of forest fires. For instance, NMR spectroscopy can be used to assess directly the pyrogenic alterations of soil substrates and vegetation and, as such, has already been widely utilized by several case studies in the context of soil quality and organic matter content (Gonzalez-Perez *et al.*, 2004; Knicker *et al.*, 2006), primarily concentrating on carbon, nitrogen and phosphate dynamics. The use of NMR spectroscopy for biological indicators is particularly valuable for recalcitrant plant components, such as the aromatic lignins or condensed biomass in general (Koegel-Knabner, 2002). Lignins show specific properties that allow an estimation of fire temperatures as well as the reconstruction of taxonomic information (Whitlock & Larsen, 2002; Knicker, 2007). On account of their natural or pyrogenic hydrophobic properties, these components remain unaltered even in aquatic archives, qualifying them for use as tracers in environmental reconstructions (Conedera *et al.*, 2009).

PROJECT AIMS AND METHODS

The principal aim of one ongoing scientific project is the assessment of fire induced erosion on Cyprus and the potential implementation of targeted mitigation measures. Because of the rather unique environmental setting and the large number of artificial sediment archives available for analysis, the project is seeking to test the applicability of black carbon for tracing soil erosion in conjunction with fire as a landscape disturbance.

Erosion assessment

The main research questions driving ongoing work by a new research project are:

- To what extent does the magnitude of soil erosion in the Troodos Mountains depend on the specific impact of forest fires?
- How can the primary parameters that define the vulnerability of the area to forest fires be measured and differentiated?

The primary task is to evaluate the vulnerability of the forested areas towards fire. In order to determine where most fires appear and to examine their impact on landscape dynamics, the establishment and maintenance of a fire database is considered essential. Historic forest fires will be registered with respect to both their spatial and temporal distribution and evaluated in a GIS environment. A geodatabase with time series of precipitation distribution, long-term climate data and fire frequencies will be established and used to undertake comparisons between these data sets.

Based on the fire geodatabase, exemplary sites of forest fires and post-fire erosion will be selected in the Troodos area for more detailed investigation. A number of ideal locations have already been located within watersheds that are connected to one of the many valley-dammed reservoirs. At these sites, all mapping results are being documented in a mobile geographical information system. The mobile mapping device (Trimble Yuma) is a rugged outdoor computer with an integrated GPS, providing a higher resolution compared to standard handheld devices. The primary tasks of these initial field surveys will be to identify and quantify sediment archives as well as to identify transport pathways and processes inside the respective catchments. Additionally, sets of soil samples will be collected from burnt and unburnt sites to determine location-specific parameters.

High resolution land-cover data that is needed for vegetation cover and fuel-load assessments will be provided by surveys of the APAESO project (Autonomous Flying Platforms for Atmospheric and Earth Surface Observations), conducted by members of the Cyprus Institute. UAS devices are used by the Cyprus Institute for a variety of scientific purposes, such as obtaining high-resolution multispectral remote sensing data to generate land-cover classifications.

Local fire reconstruction based on NMR

Due to the high number of dams (about 100) and the contrasting nature of the corresponding drainage basins, the project also aims to test the use of black carbon as an erosion indicator and as a tool for quantitative reconstruction of historic process dynamics. To this end, NMR (nuclear magnetic resonance) spectral analysis of soil organic matter will be used to draw conclusions on local fire conditions, and to trace the sediment delivery pathways down to valley-dammed reservoirs. Estimation of the long-term sediment yields of the specific catchments will be based on the gross sediment accumulation in the dams together with information on the year of construction and retention efficiency. A sediment chronology for the dam sites will be based on the collection of cores during the summer months. The data provided by the use of black carbon as an erosion tracer will be used to support catchment modelling for the eastern Mediterranean, for decision making associated with the targeting of forest fire mitigation measures.

PERSPECTIVE

The project briefly described above is still under development and its funding is currently low to non-existent. Initial field surveys will start in March 2012. The purpose of this contribution is to increase awareness of this particular application of black carbon as an erosion tracer and to provide an opportunity for wider researchers to feedback recommendations in terms of the research methodology.

Acknowledgements The authors would like to thank the reviewers for their valuable comments and recommendations for improving the manuscript. We also thank Armin Duenkeloh (University of Wuerzburg) for providing parts of his GIS data collection.

REFERENCES

Aber, J. S., Marzolff, I. & Ries, J. B. (2010) *Small-format Aerial Photography – Principles, Techniques and Geoscience Applications*. Elsevier, 268 pp.

Bakker, M. M., Govers, G., van Doorn, A., Quetier, F., Chouvardas, D. & Rounsevell, M. (2008) The response of soil erosion and sediment export to land-use change in four areas of Europe: The importance of landscape pattern. *Geomorphology* 98 (3-4), 213–226, 10.1016/j.geomorph.2006.12.027.

Bellin, N., van Wesemael, B., Meerkerk, A., Vanacker, V. & Barbera, G. G. (2009) Abandonment of soil and water conservation structures in Mediterranean ecosystems: A case study from south east Spain. *Catena* 76(2), 114–121, doi:10.1016/j.catena.2008.10.002.

Boardman, J. (2006) Soil erosion science: reflections on the limitations of current approaches. *Catena* 68(2-3), 73–86, doi:10.1016/j.catena.2006.03.007.

Boronina, A., Renard, P., Balderer, W. & Christodoulides, A. (2003) Groundwater resources in the Kouris catchment (Cyprus): data analysis and numerical modelling. *J. Hydrol.* 271(1-4), 130–149, doi:10.1016/s0022-1694(02)00322-0.

Boustras, G., Bratskas, R., Pourgouri, S., Michaelides, A., Efstathiades, A. & Katsaros, E. (2008) A Report on Forest Fires in Cyprus. *Australasian Journal of Disaster and Trauma Studies* 2.

Brown, A. G., Carey, C., Erkens, G., Fuchs, M., Hoffmann, T., Macaire, J.-J., Moldenhauer, K.-M. & Walling, D. E. (2009) From sedimentary records to sediment budgets: multiple approaches to catchment sediment flux. *Geomorphology* 108(1-2), 35–47, doi:10.1016/j.geomorph.2008.01.021.

Cammeraat, E. L. H. (2004) Scale dependent thresholds in hydrological and erosion response of a semi-arid catchment in southeast Spain. *Agriculture, Ecosystems & Environment* 104(2), 317–332, doi:10.1016/j.agee.2004.01.032.

Cerdà, A. & Doerr, S. H. (2008) The effect of ash and needle cover on surface runoff and erosion in the immediate post-fire period. *Catena* 74(3), 256–263, 10.1016/j.catena.2008.03.010.

Certini, G. (2005) Effects of fire on properties of forest soils: a review. *Oecologia* 143 (1), 1–10, 10.1007/s00442-004-1788-8.

Conedera, M., Tinner, W., Neff, C., Meurer, M., Dickens, A. F. & Krebs, P. (2009) Reconstructing past fire regimes: methods, applications, and relevance to fire management and conservation. *Quaternary Science Reviews* 28(5-6), 555–576, doi: 10.1016/j.quascirev.2008.11.005.

Cyprus Meteorological Service (2011) Meteorological Reports. Ministry of Agriculture, Natural Resources and Environment, Nicosia, Cyprus. Official website: http://www.moa.gov.cy.

Department of Forests (2011) Forest fires statistics. Ministry of Agriculture, Natural Resources and Environment, Nicosia, Cyprus. Official website: http://www.moa.gov.cy.

European Environment Agency (2009) Water resources across Europe – confronting water scarcity and drought, EEA. 2/2009.

Fox, D. M., Darboux, F. & Carrega, P. (2007) Effects of fire-induced water repellency on soil aggregate stability, splash erosion, and saturated hydraulic conductivity for different size fractions. *Hydrol. Processes* 21 (17), 2377–2384, 10.1002/hyp.6758.

Garcia-Herrera, R., Paredes, D., Trigo, R. M., Trigo, I. F., Hernandez, E., Barriopedro, D. & Mendes, M. A. (2007) The outstanding 2004/05 drought in the Iberian Peninsula: associated atmospheric circulation. *J. Hydromet.* 8(3), 483–498.

García-Ruiz, J. M. (2010) The effects of land uses on soil erosion in Spain: a review. *Catena* 81(1), 1–11, doi: 10.1016/j.catena.2010.01.001.

Gedye, S. J., Jones, R. T ., Tinner, W., Ammann, B. & Oldfield, F. (2000) The use of mineral magnetism in the reconstruction of fire history: a case study from Lago di Origlio, Swiss Alps. *Palaeogeography, Palaeoclimatology, Palaeoecology* 164(1-4), 101–110, doi:10.1016/s0031-0182(00)00178-4.

Giorgi, F. (2006) Climate change hot-spots. *Geophys. Res. Lett.* 33(8), L08707, 10.1029/2006gl025734.

Haala, N., Hastedt, H., Wolf, K., Ressl, C. & Baltrusch, S. (2010) Digital photogrammetric camera evaluation generation of digital elevation models. *Photogrammetrie – Fernerkundung – Geoinformation* 2010(2), 99–115, 10.1127/1432-8364/2010/0043.

Hadjinicolaou, P., Giannakopoulos, C., Zerefos, C., Langem, M. A., Pashiardis, S. & Lelieveld, J. (2010) Mid-21st century climate and weather extremes in Cyprus as projected by six regional climate models. *Reg. Env. Change* doi: 10.1007/s10113-010-0153-1

Hadjiparaskevas, C. (2005) Soil survey and monitoring in Cyprus. *European Soil Bureau-Research Report* 9, 97–101.

IPCC (2007) Climate Change 2007: The Physical Science Basis. In: S. Solomon, D. Qin, M. Manning, Z. Chen, M. Marquis, K. B. Averyt, M. Tignor & H. L. Miller (eds.), Contribution of Working Group I to the Fourth Assessment Report of the Intergovernmental Panel on Climate Change, p. 996. Cambridge, United Kingdom and New York, NY, USA.

Kimoto, A., Nearing, M. A., Shipitalo, M. J. & Polyakov, V. O. (2006) Multi-year tracking of sediment sources in a small agricultural watershed using rare earth elements. *Earth Surface Processes and Landforms* 31(14), 1763–1774, 10.1002/esp.1355.

Knicker, H. (2007) How does fire affect the nature and stability of soil organic nitrogen and carbon? A review. *Biogeochemistry* 85(1), 91–118, 10.1007/s10533-007-9104-4.

Koegel-Knabner, I. (2000) Analytical approaches for characterizing soil organic matter. *Organic Geochemistry* 31(7-8), 609–625, doi:10.1016/s0146-6380(00)00042-5.

Kosmas, C., Danalatos, N., Cammeraat, L. H., Chabart, M., Diamantopoulos, J., Farand, R., Gutierrez, L., Jacob, A., Marques, H., Martinez-Fernandez, J., Mizara, A., Moustakas, N., Nicolau, J. M., Oliveros, C., Pinna, G., Puddu, R., Puigdefabregas, J., Roxo, M., Simao, A., Stamou, G., Tomasi, N., Usai, D. & Vacca, A., (1997) The effect of land use on runoff and soil erosion rates under Mediterranean conditions. *Catena* 29(1), 45–59, doi:10.1016/s0341-8162(96)00062-8.

Mabit, L., Benmansour, M. & Walling, D.E. (2008) Comparative advantages and limitations of the fallout radionuclides 137Cs, 210Pbex and 7Be for assessing soil erosion and sedimentation. *J. Environmental Radioactivity* 99(12), 1799–1807, 10.1016/j.jenvrad.2008.08.009.

Michaelides, S. & Pashiardis, S. (2008) Monitoring drought in Cyprus during the 2007-2008 hydrometeorological year by using the standardized precipitation index (SPI). *European Water* 23/24, 123–131.

Milnes, E. & Renard, P. (2004) The problem of salt recycling and seawater intrusion in coastal irrigated plains: an example from the Kiti aquifer (Southern Cyprus). *J. Hydrol.* 288 (3-4), 327–343, doi: 10.1016/j.jhydrol.2003.10.010.

Morvan, X., Saby, N. P. A., Arrouays, D., Le Bas, C., Jones, R. J. A., Verheijen, F. G. A., Bellamy, P. H., Stephens, M. & Kibblewhite, M. G. (2008) Soil monitoring in Europe: A review of existing systems and requirements for harmonisation. *Sci. Total Environ.* 391(1), 1–12, 10.1016/j.scitotenv.2007.10.046.

Naveh, Z. (1975) The evolutionary significance of fire in the Mediterranean region. *Plant Ecology* 29(3), 199–208, 10.1007/bf02390011.

Naveh, Z. (2007) Fire in the Mediterranean – a landscape ecological perspective. In: *Transdisciplinary Challenges in Landscape Ecology and Restoration Ecology – Fire in Ecosystems Dynamics* (ed. by J. F. Goldammer & M. J. Jenkins), 1–20. Springer Netherlands.

Pausas, J. G. (1999) Response of plant functional types to changes in the fire regime in Mediterranean ecosystems: A simulation approach. *J. Vegetation Science* 10(5), 717–722, doi:10.2307/3237086.

Pausas, J. G. (2004) Changes in fire and climate in the Eastern Iberian Peninsula (Mediterranean Basin). *Climatic Change* 63 (3), 337–350, doi:10.1023/B:CLIM.0000018508.94901.9c.

Ries, J. B. (2010) Methodologies for soil erosion and land degradation assessment in Mediterranean-type ecosystems. *Land Degradation & Development* 21(2), 171–187, doi:10.1002/ldr.943.

Romero-Díaz, A., Alonso-Sarriá, F. & Martínez-Lloris, M. (2007) Erosion rates obtained from check-dam sedimentation (SE Spain). A multi-method comparison. *Catena* 71(1), 172–178, doi: 10.1016/j.catena.2006.05.011.

Seeger, M. (2007) Uncertainty of factors determining runoff and erosion processes as quantified by rainfall simulations. *Catena* 71(1), 56–67, doi:10.1016/j.catena.2006.10.005.

Shakesby, R. A. & Doerr, S. H. (2006) Wildfire as a hydrological and geomorphological agent. *Earth-Science Reviews* 74(3-4), 269–307, doi:10.1016/j.earscirev.2005.10.006.

Sirvent, J., Desir, G., Gutierrez, M., Sancho, C. & Benito, G. (1997) Erosion rates in badland areas recorded by collectors, erosion pins and profilometer techniques (Ebro Basin, NE-Spain). *Geomorphology* 18(2), 61–75, doi: 10.1016/s0169-555x(96)00023-2.

Varga, R. J. & Moores, E. M. (1985) Spreading structure of the Troodos ophiolite, Cyprus. *Geology* 13(12), 846–850, 10.1130/0091-7613(1985)13<846:ssotto>2.0.co;2.

Ventura, E., Nearing, M. A., Amore, E. & Norton, L. D. (2002) The study of detachment and deposition on a hillslope using a magnetic tracer. *Catena* 48(3), 149–161, doi: 10.1016/s0341-8162(02)00003-6.

Whitlock, C. & Larsen, C. (2002) Charcoal as a fire proxy. In: *Tracking Environmental Change Using Lake Sediments 03; Terrestial, Algal, and Siliceous Indicators,*(ed. by J. P. Smol, H. J. B. Birks & W. M. Last), 75–97. Kluwer Academic Publishers, Dordrecht, Netherlands.

Suspended sediment yield following wildfires in a mixed species eucalypt forest, southeastern Australia

DEIRDRE DRAGOVICH[1], ASHLEY A. WEBB[2] & REZA JAMSHIDI[1]

1 *School of Geosciences F09, University of Sydney, Sydney 2006, Australia*
deirdre.dragovich@sydney.edu.au

2 *Forests NSW; PO Box 4019, Coffs Harbour Jetty NSW 2450, Australia*

Abstract In June 2001, flow and suspended sediment monitoring equipment was installed at the outlets of four study sub-catchments in Kangaroo River State Forest, southeastern Australia. A moderate severity wildfire in October 2001 burnt 94% of a 443-ha sub-catchment and the same wildfire, combined with an earlier fire in August 2001, burnt through 61% of the adjacent 367-ha sub-catchment. Neither of the remaining sub-catchments experienced these wildfires. The 2001 wildfires occurred during a drought year which was followed by an extended period of low rainfall. Substantial flows did not occur in either of the burnt sub-catchments until February 2003. As a consequence, suspended sediment yields in the immediate post-fire period were minimal and not significantly different in the burnt and unburnt sub-catchments. A substantial sediment pulse was generated during the summer rains of February (223.7 mm) and March (200.5 mm) 2003 in all sub-catchments, with sediment responses being similar in the burnt and unburnt areas. This case study illustrates the importance of the timing and magnitude of post-fire rainfall events in determining the likelihood of significant sediment transport following wildfires.

Key words wildfire; drought; catchment response; Kangaroo River; southeastern Australia

INTRODUCTION

The rate of sediment transfer commonly increases following wildfires (Zierholz *et al.*, 1995; Prosser & Williams, 1998), especially when high-intensity rainfall events occur before vegetation cover has become re-established (Smith & Dragovich, 2008). The sediment response to post-fire conditions is also dependent on fire severity (Prosser, 1990; Dragovich & Morris, 2002) with low severity and prescribed burns often resulting in only minor erosion. Regardless of fire severity, any high intensity rainfall event occurring post-fire is likely to generate accelerated sediment redistribution. In the higher rainfall areas of Australia, fuel loads accumulate during wetter years and these areas are prone to wildfires during extremely hot summers and/or drought periods. Against this context, this study investigated suspended sediment yields from adjacent sub-catchments that had been partly burnt, almost completely burnt, or were entirely unburnt, with the aim of identifying potential differences in fire-generated accelerated erosion related to the spatial extent of burnt areas.

STUDY AREA

The study was carried out in Kangaroo River State Forest in southeastern Australia. Monitoring equipment was installed in four adjacent sub-catchments (two sets of paired sub-catchments). The terrain is dissected by V-shaped valleys with minimal development of flood plains in the two northern sub-catchments, but small pockets of flood plain deposits and benches exist in the larger southern sub-catchments. In most areas the generally steeply-sloping terrain supports a dense understory and mixed species eucalypt forests dominated by Spotted Gum (*Corymbia maculata*), Sydney Blue Gum (*Eucalyptus salinga*), Blackbutt (*E. pilularis*) and Flooded Gum (*E. grandis*), along with stands of Grey Ironbark (*E. paniculata*) and Grey Gum (*E. propinqua*) on summits and upper slopes. On wet lower slopes, White and Red Mahogany (*E. acmenoides* and *E. resinifera*) occur, especially in shaded locations (Forestry Commission of NSW, 1989). Prior to 2001, selective harvesting for saw-logs took place before 1993 in the two southern (unburnt) sub-catchments and some harvesting probably occurred in the northern (burnt) sub-catchments in the 1970s.

Local soils are dominated by the Black Mountain soil landscape which is characterised by well-drained structured Yellow Earths and Brown Earths (Milford, 1996), often occurring in deep (>150 cm) units. Milford (1996) categorised the Black Mountain and most other soil landscapes in the study sub-catchments as having high to very high erodibility, with only relatively small areas described as having low to moderate erodibility.

The climate is subtropical, with maximum rainfall occurring during summer when thunderstorm activity is common. Average annual rainfall at Coffs Harbour (Australian Bureau of Meteorology climate station 059040), about 40 km southeast of the study area, is 1693 mm. Average monthly temperature is 13.2°C in July and 23.2°C in January and February. Since the commencement of the study in 2001, and up to 2009, annual rainfall at the raingauge located in the southwest of the sub-catchments has averaged about 1400 mm. The years included in this study (June 2001 to June 2004) generally experienced low annual rainfall, with totals for the 12 months from June to May in each year being 643 mm to May 2002, 958 mm to May 2003, and 1032 mm to May 2004.

METHODS AND MATERIALS

Flow and suspended sediment gauges were installed at the outlets of four adjacent sub-catchments. Two sub-catchment "pairs" were designed to assess the impacts of selective logging, each pair comprising a "Control" and an "Impact" sub-catchment. Following the installation of gauges in June 2001, a low to moderate severity wildfire in October 2001 burnt 94% of a 443-ha sub-catchment called Impact 1 (I-1). The same wildfire, combined with an earlier fire in August 2001, burnt through 61% of the adjacent 367-ha sub-catchment called Control 1 (C-1) (Fig. 1). The 2001 fires were the most recent to have affected these two northern sub-catchments (C-1 and I-1). Ten near-surface soil samples collected in partly burnt C-1 contained varying amounts of charcoal, but the presence of charcoal, which is readily transported, was not confined to the burnt areas. The two southern sub-catchments called Control 2 (C-2) and Impact 2 (I-2) were not burnt in the 2001 fires.

Streamgauging stations were installed at the outlet of each sub-catchment, while both a tipping bucket (pluviometer) and manual raingauge were installed and maintained at a central location (Fig. 1). Streamgauges were located on bedrock controls upstream of tributary junctions on each stream and in similar geomorphic settings. Each station was instrumented with an automatic pump water sampler (ISCO 3700), a datalogger (Datataker DT50), pressure transducer and staff gauge, powered by 12V batteries charged by a solar panel. Stream height was logged at six-minute intervals and converted to discharge using rating curves derived from velocity–area gaugings undertaken at a range of flows. Water samples, 500 mL in volume, were automatically pumped from each stream by a stage-activated sampler (ISCO 3700 model) throughout flood events, on the rising and falling limbs of the hydrograph. In addition, weekly water samples were pumped from each stream during periods of baseflow. At each data download visit (fortnightly or more frequently during floods) water samples were retrieved from the automatic samplers, refrigerated and couriered to the laboratory where they were analysed for turbidity and suspended sediment concentration according to standard methods (APHA, 1998). The comprehensive flood and baseflow sampling facilitated accurate estimates of suspended sediment loads.

Rainfall figures for all comparisons were those recorded at the gauging station installed in the southwestern part of the study area in 2001. Initial comparisons were made for streamflow and sediment yields between sub-catchments C-1 (61% burnt) and I-1 (94% burnt) (Fig. 1). Subsequently both burnt sub-catchments 1 (C-1 and I-1) were compared with unburnt sub-catchments 2 (C-2 and I-2). These records were investigated for the period from June 2001 to June 2004, which included an initial low rainfall period and two high rainfall events (February–March 2003 and January–March 2004).

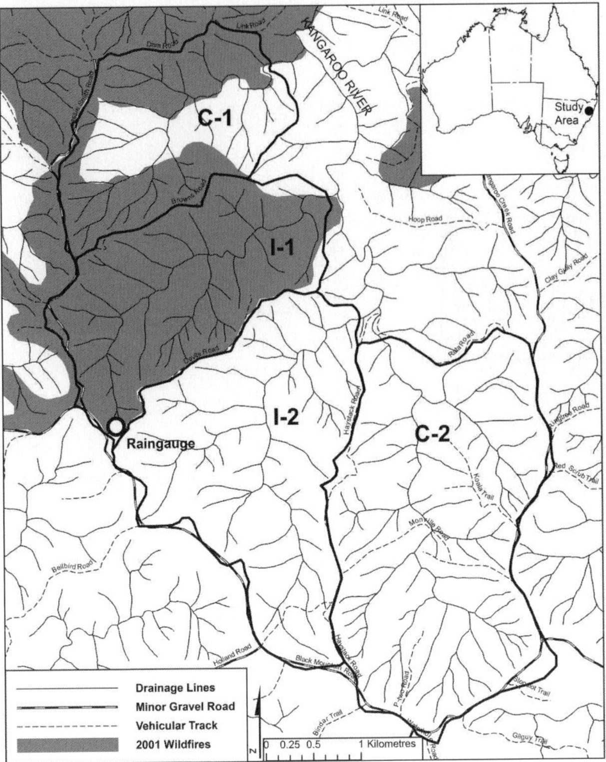

Fig. 1 Extent of the 2001 wildfires in the study sub-catchments in Kangaroo River State Forest.

RESULTS

Partly burnt (C-1) and burnt (I-1) sub-catchments

Monthly rainfall exceeded 100 mm on only three occasions between June 2001 and January 2003, giving an average monthly rainfall for this period of 49.6 mm compared with the long-term average at Coffs Harbour of 141.1 mm. Streamflow was negligible or absent in both sub-catchments over this period: C-1 recorded 10 months with zero flow months, and I-1, 9 months. In February and March 2003, rainfall totalled 223.7 mm and 200.5 mm, respectively, and these falls generated substantial streamflow in both sub-catchments. Streamflow was greater in the partly burnt (C-1) sub-catchment than in the nearly completely burnt (I-1) one. Subsequent minor peaks in rainfall and streamflow occurred in May 2003 and January 2004, with a second major peak in March 2004. The high rainfall in January 2004 (290 mm) followed a relatively dry period of several months and streamflow did not respond substantially until further heavy rain fell (256 mm) in March 2004 (Fig. 2).

Given the almost complete absence of streamflow until February–March 2003, recorded sediment yields over that period were very low. Yields were similar in both sub-catchments for these minor rainfall events. In March 2003, yields were greater in the partly burnt (C-1) than in the burnt (I-1) sub-catchment, and the same pattern between the sub-catchments was evident in the second high rainfall period of January–March 2004 (Fig. 2).

Streamflow and sediment yield were thus both greater for the partly burnt (C-1) than for the burnt (I-1) sub-catchment (Fig. 3). Cumulative streamflow patterns for sub-catchments C-1 and I-1 were similar, although the partly burnt area (C-1) had consistently higher values than the burnt (I-1) sub-catchment (Fig. 4). The two substantial rainfall events over the study period were reflected in steep, short-term increases in cumulative streamflow.

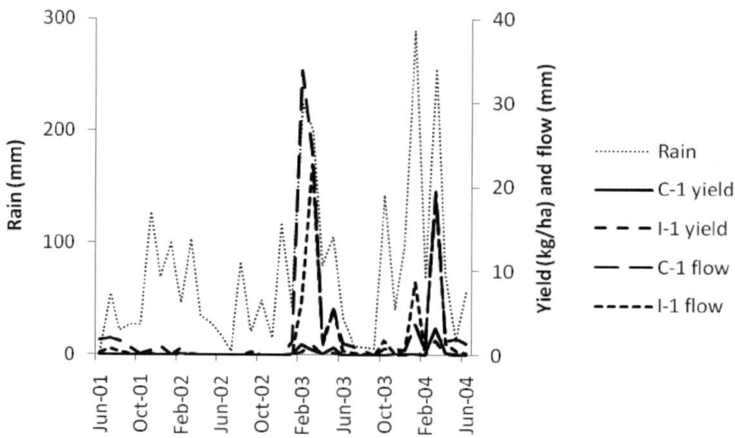

Fig. 2 Rainfall, sediment yield and streamflow in the partly burnt (C-1) and burnt (I-1) sub-catchments.

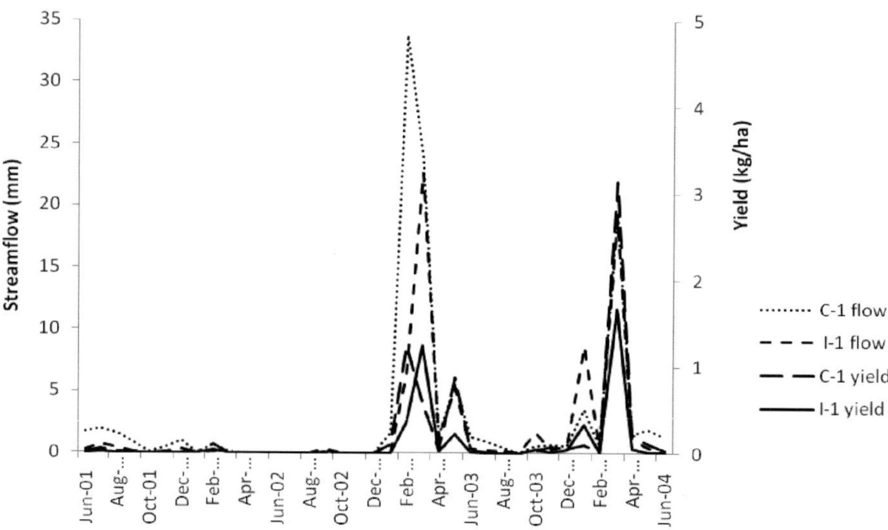

Fig. 3 Streamflow and sediment yield in the partly burnt (C-1) and burnt (I-1) sub-catchments.

Burnt sub-catchments C-1 and I-1 (Sub-Cs1), and unburnt sub-catchments C-2 and I-2 (Sub-Cs2)

The overall pattern of rainfall and streamflow for the burnt and unburnt sub-catchments was dominated by the substantial rainfalls in February–March 2003 and January–March 2004. Although streamflow in the unburnt sub-catchments (Sub-Cs2) recorded more than double the rates in the burnt sub-catchments (Sub-Cs1) in the 2004 heavy rains, the high March 2004 sediment yields were similar for both sets of sub-catchments (Fig. 5). Until that event the unburnt sub-catchments (Sub-Cs2) generally registered higher sediment yields. Cumulative yields for both the burnt and unburnt sub-catchments showed marked increases in response to the major rainfall events in 2003 and 2004, with the unburnt sub-catchments having higher streamflows and higher sediment yields (Fig. 6).

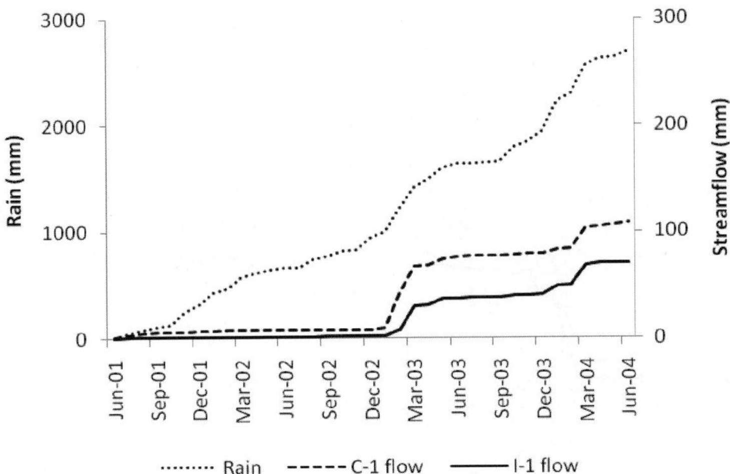

Fig. 4 Cumulative rainfall and streamflow in the partly burnt (C-1) and burnt (I-1) sub-catchments.

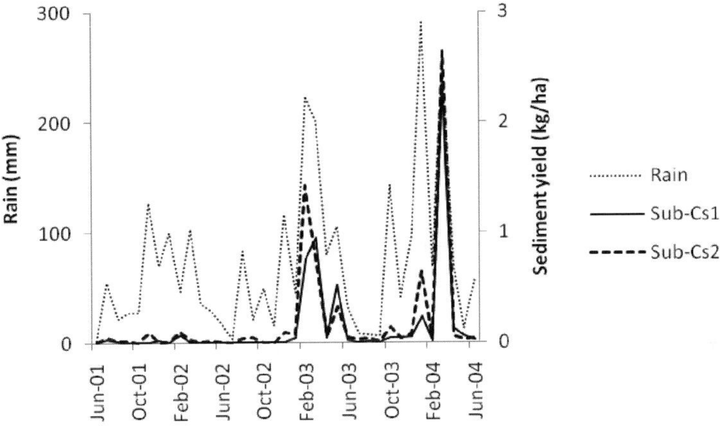

Fig. 5 Rainfall and sediment yield in burnt sub-catchments C-1 and I-1 (Sub-Cs1), and unburnt sub-catchments C-2 and I-2 (Sub-Cs2).

DISCUSSION

Burnt areas are expected to generate accelerated runoff to channels and greater suspended sediment loads (e.g. Moody & Martin, 2001; Lane *et al.*, 2006; Reneau *et al.*, 2007; Sheridan *et al.*, 2007). Such post-fire sediment may be delivered to streams by slope wash and/or by erosion of channel beds or banks during higher flows. More extensively burnt areas could be expected to produce greater runoff and sediment transport than little affected or unburnt catchments. In the February–March 2003 rains in this study, streamflow was greater in partly burnt sub-catchment C-1 than in nearly completely burnt I-1 (two-month totals of 57.8 mm and 29.0 mm, respectively). Sediment yields were similar for both sub-catchments (two-month totals of 1.8 and 1.6 kg/ha, respectively). In the heavy rainfalls of March 2004, streamflow in the two sub-catchments was similar (19.5 and 19.0 mm). However, sub-catchment C-1 with 61% of its area burnt in wildfires in 2001, generated considerably higher sediment yields than adjacent sub-catchment I-1 with 94% of its area burnt (3.14 kg/ha compared with 1.67 kg/ha). This response may be partly attributable

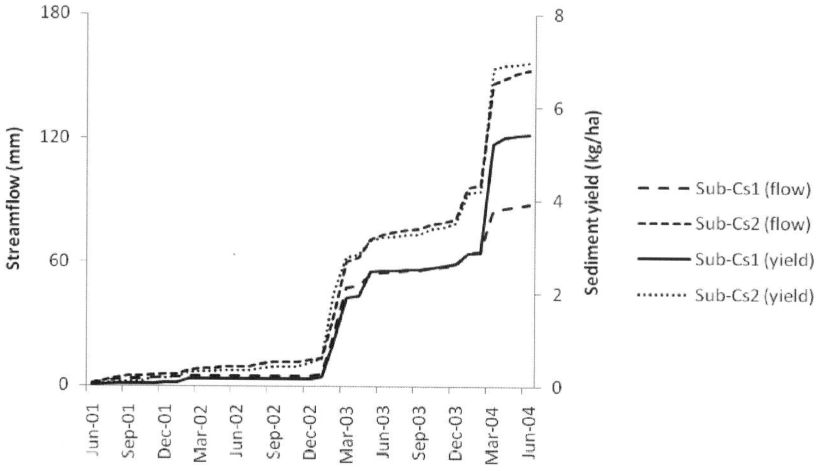

Fig. 6 Cumulative sediment yield and streamflow in the burnt sub-catchments C-1 and I-1 (Sub-Cs1), and unburnt sub-catchments C-2 and I-2 (Sub-Cs2).

to the low to moderate severity wildfires and to contrasts in sub-catchment attributes which could produce varying results even in the absence of wildfire as a major landscape disturbance.

Riparian vegetation is dense in all the sub-catchments. The vegetation cover *C* values for RUSLE were estimated for 2011 using NDVI-based analysis of vegetation (Jamshidi *et al.*, 2012). Low *C* values occurred along stream lines reflecting the dense vegetation cover in those portions of the study area. NDVI analysis of the *C* factor was applied to 2005 images and mean *C* values for the partly burnt (C-1) and burnt (I-1) sub-catchments were estimated at 0.0034 and 0.0032, respectively. Mean *C* values for both unburnt sub-catchments (C-2 and I-2) were lower (0.0021 and 0.0025). Riparian vegetation could be expected to intercept hillslope sediment delivery in sub-catchment C-1, as this was the unburnt area in the 2001 fires. Despite this, sub-catchment C-1 had higher sediment yields than sub-catchment I-1. In addition to the importance of vegetation cover is the presence of a litter layer, which may be capable of providing substantial moisture storage for post-fire rainfalls (Leighton-Boyce *et al.*, 2007), thereby reducing overland flow. Previous work has suggested that augmented flows following heavy rains lead to greater mobilisation of channel bed/bank deposits (Moody & Martin, 2009) and this may have contributed to the sharply increased sediment yields noted in the two post-fire high rainfall events that were apparent in both the burnt and unburnt sub-catchments. Sheridan *et al.* (2011) proposed that sediment sources shift from being slope-dominated to channel-dominated as vegetation recovers in the years following fires. Accelerated amounts of hillslope-derived sediments are transferred post-fire, but not necessarily transported into channels (Shakesby *et al.*, 2007); temporary sediment sinks on hillslopes probably become increasingly important as fire severity diminishes.

Sediment transfer at the plot scale cannot be readily translated to sediment yield at catchment scales (Lane *et al.*, 2006), but sediment response patterns to post-fire rainfall events are similar at the two scales. Using runoff troughs to capture mobilised sediment, Prosser & Williams (1998) found that sediment yields for burnt plots increased markedly with high intensity rainfall, results that were consistent with earlier rainfall and plot data on post-fire sediment transfer reported by Blong *et al.* (1982) and Atkinson (1984), and observations by Zierholz *et al.* (1995). Nyman *et al.* (2011) found that high intensity storms following wildfires in small (<100 ha) and steep (25° to 35°) catchments generated considerable amounts of sheet erosion which made a large contribution to suspended sediment yield, and Lane *et al.* (2006) also noted that one or two heavy thunder-storms were responsible for mobilising and transporting most of the post-fire sediment load. The impacts of low intensity rainfalls, which encourage vegetation regrowth following fire and are associated with minimal erosion and negligible flows (Condina *et al.*, 1984; Chessman, 1986), are less frequently reported. These conditions were observed during this study in the Kangaroo River State Forest.

During this study, the combined unburnt sub-catchments (C-2 and I-2: called Sub-Cs2) generally recorded higher streamflows and sediment yields than the burnt sub-catchments (C-1 and I-1: called Sub-Cs1). Drainage density was a potential variable influencing these yields. However, drainage density was greatest in the partly burnt (C-1) (5.035 km/km^2) and lowest in the burnt (I-1) (4.137 km/km^2) sub-catchment. Corresponding values for the unburnt sub-catchments (C-2 and I-2) were 4.548 km/km^2 and 4.251 km/km^2, respectively. Based on 10-m pixels in a DEM, the proportion of each sub-catchment covered by slopes exceeding 15° was nearly 65% for both partly burnt C-1 and unburnt C-2, compared with over 70% for both I-1 (burnt) and I-2 (unburnt). Slopes greater than 30° covered between 8% and 10% of I-1 and I-2, respectively. Of the four sub-catchments, burnt I-1 had the lowest drainage density and the greatest proportion of slopes exceeding 15°, while partly burnt C-1 had the highest drainage density and lowest proportion of steeper slopes. The contribution of these terrain characteristics to observed sediment yield differences is uncertain as the extent to which slope steepness is offset by drainage density is not known.

Like many hydrological studies focused on landscape disturbance, this experiment was not able to assemble lengthy pre-fire flow and sediment yield records due to the short period between the installation of monitoring equipment and the wildfires. A longer pre-fire record would have allowed for robust comparisons between pre- and post-fire responses in both burnt and unburnt sub-catchments, as was possible for paired catchments in Victoria, Australia (Lane *et al.*, 2006; Bren, 2012). The lack of robust baseline data is a problem for many paired catchment studies across the world. A further constraint on interpreting the observed differences between the burnt and unburnt areas used in this study was the likelihood of rainfall variations over the individual sub-catchments, especially relating to the incidence and intensity of highly localised summer thunderstorms.

CONCLUSION

This study investigated possible differences in sediment yield responses between partly burnt, almost completely burnt, and unburnt sub-catchments. The expectation of greater sediment yields in sub-catchments most affected by wildfire was not observed. Although part of the reason for this may have been topographic, hydrological or soil differences between the individual sub-catchments combined with the low to moderate intensity of the fires, the intervening period between equipment installation in June 2001 and the February–March 2003 rain event included only six (non-consecutive) months with rainfall of less than 25 mm. This consistent presence of moisture delivered in low intensity falls in the 20 months following fires allowed for the re-growth of forest groundcover and produced subdued streamflow and sediment yield responses to wildfire. The results of this study therefore suggest that the timing and intensity of post-fire rainfall events are important determinants of erosion and sediment transport responses to wildfires.

Acknowledgements The Kangaroo River paired catchment study was funded by Forests NSW. The authors are grateful to Lisa Turner, Brad Jarrett, Dennis Burt, Ian Hanson, John Major and Geoff Heagney for providing field and laboratory assistance.

REFERENCES

APHA (1998) *Standard Methods for the Examination of Water and Wastewater*, 20th edition. American Public Health Association, Washington DC.

Atkinson, G. (1984) Erosion damage following bushfires. *J. Soil Conserv. Service NSW* 40, 4–9.

Blong, R. J., Riley, S. J. & Crozier, P. J. (1982) Sediment yield from runoff plots following bushfire near Narrabeen Lagoon, NSW. *Search* 13, 36–38.

Bren, L. (2012) Hydrologic impact of fire on the Croppers Creek paired catchment experiment. In: *Revisiting Experimental Catchment Studies in Forest Hydrology* (ed. by A. A. Webb), 154–165. IAHS Publ. 353. IAHS Press, Wallingford, UK.

Chessman, B. C. (1986) Impact of the 1983 wildfires on river water quality in East Gippsland, Victoria. *Aust. J. Mar. Freshw. Res*. 37, 399–420.

Dragovich, D. & Morris, R. (2002) Fire intensity, run-off and sediment movement in eucalypt forest near Sydney, Australia. In: *Applied Geomorphology* (ed. by R. J. Allison), 145–164. John Wiley & Sons, Chichester, England.

Forestry Commission of NSW (1989) Forest types in New South Wales. Research Note 17, Forestry Commission of NSW, Sydney. 95 pp.

Jamshidi, R., Dragovich, D. & Webb, A. A. (2012) Native forest *C* factor determination using satellite imagery in four sub-catchments. In: *Revisiting Experimental Catchment Studies in Forest Hydrology* (ed. by A. A. Webb *et al.*), 64–73. IAHS Publ. 353. IAHS Press, Wallingford, UK.

Lane, P. N. J., Sheridan, G. J. & Noske, P. J. (2006) Changes in sediment loads and discharge from small mountain catchments following wildfire in south eastern Australia. *J. Hydrol.* 331, 495–510.

Leighton-Boyce, G., Doerr, S. H., Shakesby, R. A. & Walsh, R. P. D. (2007) Quantifying the impact of soil water repellency on overland flow generation and erosion: a new approach using rainfall simulation and wetting agent on *in situ* soil. *Hydrol. Processes* 21, 2337–2345.

Milford, H. B. (1996) Soil Landscapes of the Dorrigo 1:100 000 Sheet Map. Dept. of Land and Water Conservation, Sydney, NSW. Report. 252 pp.

Moody, J. A. & Martin, D. A. (2001) Initial hydrologic and geomorphic response following a wildfire in the Colorado Front Range. *Earth Surf. Processes Landf.* 26, 1049–1070.

Moody, J. A. & Martin, D. A. (2009) Synthesis of sediment yields after wildfire in different rainfall regimes in the western United States. *Int. J. Wildland Fire* 18, 96–115.

Nyman, P., Sheridan, G. J., Smith, H. G. & Lane, P. N. J. (2011) Evidence of debris flow occurrence after wildfire in upland catchments of south-east Australia. *Geomorphology* 125, 383–401, doi:10.1016/j.geomorph.2010.10.016.

Prosser, I. (1990) Fire, humans and degradation at Wangrah Creek, Southern Tablelands, NSW. *Aust. Geogr. Studies* 28, 77–95.

Prosser, I. P. & Williams, L. (1998) The effect of wildfire on runoff and erosion in native *Eucalyptus* forest. *Hydrol. Processes* 12, 251–265.

Reneau, S. L., Katzman, D., Kuyumjian, G. A., Lavine, A. & Malmon, D. V. (2007) Sediment delivery after a wildfire. *Geology* 35, 151–154.

Shakesby, G. J., Wallbrink, P. J., Doerr, S. H., English, P. M., Chafer, C. J., Humphreys, G. S., Blake, W. H. & Tomkins, K. M. (2007) Distinctiveness of wildfire effects on soil erosion in south-east Australian eucalypt forests assessed in a global context. *Forest Ecol. Management* 238, 347–364.

Sheridan, G. J., Lane, P. N. J. & Noske, P. J. (2007) Quantification of hillslope runoff and erosion processes before and after wildfire in a wet *Eucalyptus* forest. *J. Hydrol*. 343, 12–28.

Sheridan, G. J., Lane, P. N. J., Sherwin, C. B. & Noske, P. J. (2011) Post-fire changes in sediment rating curves in a wet *Eucalyptus* forest in SE Australia. *J. Hydrol.* 409, 183–195.

Smith, H. G. & Dragovich, D. (2008) Post-fire hillslope erosion response in a sub-alpine environment, south eastern Australia. *Catena* 73, 274–285, doi:10.1016/j.catena.2007.11.2003

Zierholz, C., Hairsine, P. & Booker, F. (1995) Run-off and soil erosion in bushland following the Sydney bushfires. *Aust. J. Soil and Water Conserv.* 8, 28–37.

Sediment yields and water quality effects of severe wildfires in southern British Columbia

PETER JORDAN
British Columbia Ministry of Forests, Lands & Natural Resource Operations, 1907 Ridgewood Road, Nelson, British Columbia V1L 6K1, Canada
peter.jordan@gov.bc.ca

Abstract Following wildfire, significant erosion and water quality impacts can occur. However, in British Columbia (BC), Canada, such impacts have seldom been reported. In 2007, several large wildfires occurred in southeastern BC, some in community watersheds. Research sites were established at three fire locations, and plot-scale measurements were made of erosion using silt-fence sediment traps. Watershed-scale measurements of runoff, sediment yield and chemical water quality were made on the Sitkum fire, which burned 39% of a community watershed. A nearby research watershed provided a comparison. Significant surface erosion occurred in some burned areas, but watershed-scale sediment yield increased only slightly, as little sediment reached stream channels. Nitrate levels were elevated after the fire, but were well within accepted limits for drinking water quality. The minimal effects on physical water quality are probably due to low rainfall intensities, the nival runoff regime, and low connectivity between slopes and stream channels in a glaciated landscape.

Key words post-wildfire erosion; wildfire; sediment yield; turbidity; water quality; nitrate; British Columbia, Canada

INTRODUCTION

The incidence of large, severe wildfires in southern British Columbia (BC), Canada, has apparently been increasing in the last several decades, most dramatically with the extreme fire season of 2003 (Filmon *et al.*, 2004). After the 2003 fires, floods and landslides caused severe damage to property, infrastructure, and stream channels below five fires (Jordan & Covert, 2009). Such events had rarely or never been documented previously in Canada, although similar post-wildfire events have often been observed elsewhere (e.g. Shakesby & Doerr, 2006). In 2007, 2009, and 2010, severe wildfire seasons again occurred in parts of southern BC, with many fires burning near populated areas, including watersheds used for community and domestic water supply.

Although there is often concern that drinking water quality will be adversely affected by wildfires, there have been few reports of significant impacts on water quality in BC, other than minor increases in turbidity in the first year after the fires. The exception to this general observation is the few stream channels in which large debris flows or debris floods have occurred (Jordan & Covert, 2009).

In 2007, the BC Forest Service began a research project to investigate post-wildfire natural hazards. Several large wildfires occurred in 2007 near populated areas in southeastern BC, which provided opportunities to study the effects of fires on the hazards of landslides, erosion and flooding. The project included measurements of soil erosion at study plots on four fires, and watershed-scale measurements of streamflow, sediment yield and water quality sampling on one fire. In this paper, the results of these measurements are discussed in the context of drinking water quality.

PHYSICAL WATER QUALITY IMPACTS OF WILDFIRE

The most frequently reported effect of wildfire on water quality is increased sediment concentration. In many studies from the USA, substantial increases in soil erosion, turbidity, and suspended sediment concentration following wildfires have been documented (Beschta, 1990; Robichaud *et al.*, 2000; Neary *et al.*, 2005; Moody & Martin, 2009). In most cases, the source of sediment is soil erosion in burned areas, the amount of which depends on soil burn severity, the degree of water repellency and the rainfall intensities that occur after burning (Robichaud *et al.*,

2000; Doerr *et al.*, 2006). An increase in peak streamflow after a fire can entrain additional sediment from erosion of channel banks, and an increased incidence of debris flows in susceptible terrain can also cause high suspended sediment concentrations downstream (Beschta, 1990).

In western Canada, three studies have examined the effects of large wildfires on the water quality of streams. The first, reported by Gluns & Toews (1989), sampled water quality for three years following a wildfire in Matthew Creek in southeastern BC, a community watershed supplying drinking water to the city of Kimberley. The fire burned the lower portions of two tributaries of Matthew Creek, and samples were collected at points upstream and downstream of the burn on these two streams and at corresponding locations on an adjacent unburned tributary. Gluns & Toews (1989) did not observe any significant increase in turbidity, although they noted that their sampling schedule (biweekly during the spring freshet period) was unsuitable for detecting changes in turbidity or suspended sediment responses, which are transient in nature and characterized by high temporal variability. Secondly, a study of water quality after a large wildfire in the Crowsnest Pass area of southwestern Alberta (Bladon *et al.*, 2008; Silins *et al.*, 2009) collected samples in seven watersheds of similar size, including five which were entirely or partly burned, and two unburned reference watersheds adjacent to the fire. They collected daily samples using automatic pump samplers, and observed suspended sediment concentrations ranging from 6 to 15 times greater in the burned, compared to the unburned watersheds. Thirdly, following the 2003 McClure fire near Barriere, BC, Eaton *et al.* (2010) monitored streamflow and suspended sediment concentration for four years in Fishtrap Creek, a watershed which was 62% burned, and a nearby similar but unburned watershed. They did not observe any difference in suspended sediment response.

CHEMICAL WATER QUALITY IMPACTS OF WILDFIRE

Following wildfire, concentrations of some solutes in streamwater can increase as a result of several processes, including loss of vegetation and the resulting reduced uptake of nutrients from the soil, leaching of solutes from ash and increased erosion of the burned soil by surface runoff (Beschta, 1990; Neary *et al.*, 2005). Numerous studies have shown that increased nutrient concentrations in water can follow forest disturbances including harvesting, prescribed burning and wildfire, with the effects generally increasing in that order (Neary *et al.,* 2005; Pike *et al.*, 2010). Higher burn severity is likely to lead to greater water quality effects, due to greater consumption of organic material and corresponding increased soil erodibility (Pike *et al.*, 2010).

Nitrogen is the nutrient which usually shows the greatest increase after fire. Nitrate (NO_3) is the form of nitrogen of most interest, as it is more mobile and typically occurs in much greater concentrations than NO_2, NH_4 and dissolved organic nitrogen (Neary *et al.*, 2005; Bladon *et al.*, 2008). Substantial increases in NO_3 concentrations have often been observed following wildfire, but these have rarely exceeded drinking water quality guidelines (Neary *et al.*, 2005; Smith *et al.*, 2011). Other solutes that have been found to increase in some studies include PO_4, SO_4 and organic carbon (Neary *et al.*, 2005; Pike *et al.*, 2010).

At Matthew Creek, Gluns & Toews (1989) measured a large increase in NO_3 concentration in the burned watersheds, which was greatest in the second year after the fire. Lesser, but significant, increases in other parameters were observed, including PO_4, total alkalinity and total hardness. Following the Crowsnest Pass wildfire, Bladon *et al.* (2008) reported increases in the concentration of various forms of nitrogen in burned areas, ranging from 1.5 to 6.5 times those observed in the unburned watersheds. The increase was greatest in the first year after the fire, and declined for the following three years. A similar increase occurred for phosphorus (Silins *et al.*, 2009), with the greatest increase observed in the second year after the fire. In neither of these studies, however, was the drinking water quality guideline for nitrate (10 mg/L NO_3 as N; BC Ministry of Environment, 2009) exceeded. There is no published drinking water guideline for phosphorus in either BC or Alberta, but a suggested federal-provincial guideline is 0.2 mg/L (Gluns & Toews, 1989) and this was not exceeded in either study following the respective wildfires.

STUDY SITES AND METHODS

Following the 2007 wildfires in southeastern BC, study sites were located in the burned area of three fires: the Sitkum fire near Nelson; the Springer fire near Slocan; and the Pend d'Oreille fire near Trail (Fig. 1). In 2009, a study site was added in the area burned by the Terrace Mountain fire near Kelowna, with the objective of including a lower elevation, drier, forest type. The respective areas burned by these fires were approximately 3100, 1100, 4000 and 9300 ha, and they included a substantial proportion of high burn severity. The study sites covered only a small portion of each fire; they were chosen on the basis of access, high burn severity and areas which presented high risks to downslope or downstream values, such as public safety or community water supplies. Some summary information for each study site is given in Table 1.

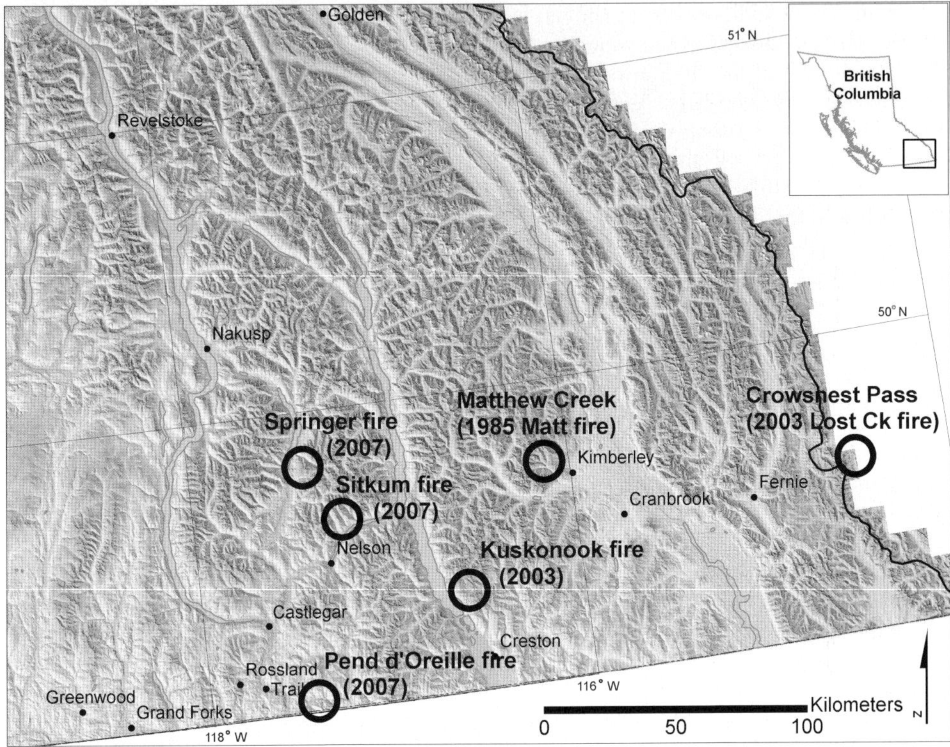

Fig. 1 Location map of southeastern British Columbia, showing the wildfires referred to in this paper. (The Terrace Mountain fire is to the west of this map.)

Table 1 Site characteristics and data from the soil erosion plots for each year after each fire.

Fire and year[1]	Elevation	Aspect	Soil texture[2]	Precipitation (mm)[3]	Mean sediment yield from erosion plots (Mg/ha); years after fire				
					n[4]	0	1	2	3
Springer (2007)	1465–1560	SW	SL	970	6	0.44	4.15	0.69	0.22
Sitkum (2007)	1600–1800	NE	SL	1290	2		0.58	6.36	1.28
Pend d'Oreille (2007)	1145–1330	SW	SiL	880	4		1.08	0.14	0
Terrace Mtn (2009)	1000–1100	W	SL	550	3	0.70	0.31	0.02	

1. Site data are for the vicinity of plots, not for the entire fire.
2. SL = sandy loam; SiL = silty loam.
3. Mean annual precipitation estimated from PRISM climate data grid (Spittlehouse, 2006).
4. Number of untreated erosion plots in each fire. All plots were in areas of high soil burn severity.

In each study area, silt fence erosion plots were installed to measure soil erosion in burned areas. Other installations in the study sites included rainfall simulation experiments to study soil erosion under high rainfall intensity conditions (Covert & Jordan, 2009), a network of raingauges, and transects to sample soil moisture, soil burn severity and water repellency.

The silt fence plots used the design of Robichaud & Brown (2002), and were installed for the original purpose of evaluating the effectiveness of mulch treatments which were applied on portions of two fire impacted slopes. In this paper, only the data from untreated plots are reported. Each silt fence plot was 3–5 m wide, and 12–16 m long. Hillslope angles on which they were installed ranged from 31 to 90%. Several times each season, including after the snowmelt period and after each major rainfall event, sediment was removed from the silt fences and weighed, and a sample was returned to the laboratory for dry weight measurement and further analysis.

The Sitkum fire burned 39% of the drainage area of Sitkum Creek, a community watershed supplying about 25 homes on an alluvial fan on Kootenay Lake. This fire was therefore considered to be high risk with respect to water quality and potential flood damage. A stream gauging station was established at the fan apex in April 2008, and an Isco 6700 automatic pump sampler was installed and operated from April to October for four years after the fire. It was programmed to take daily samples during the spring freshet period, and every three days at other times. Samples were analysed for turbidity using a Hach 2100N turbidimeter, and for suspended sediment concentration if they exceeded a threshold turbidity of 1 NTU.

A nearby stream, Redfish Creek, is very similar to Sitkum Creek in drainage area, topography, geology and soils. It has been a BC Forest Service research watershed since 1992, and at the time of the fire in 2007, it had 34 years of streamflow data and 15 years of turbidity and sediment data. It was used as a reference watershed to investigate the effects of the fire on streamflow, suspended sediment and water quality in Sitkum Creek. As forest fires cannot be planned, it is rarely possible to set up a controlled experiment to investigate such changes. It cannot be established with any certainty that pre-fire hydrological conditions in the two watersheds were the same, but in this case we believe that they are reasonably similar. Figure 2 provides a map of the two watersheds, based on a satellite image.

Redfish Creek and Sitkum Creek have similar drainage areas, 27.2 and 27.0 km^2 respectively. Both Sitkum and Redfish creeks are underlain by granite of the Nelson batholith, and have similar soils derived from glacial till and colluvium of granitic origin. Redfish Creek has several small lakes, which Sitkum Creek lacks. They drain mountainous terrain, and above the main valley bottoms most precipitation falls as snow. The runoff regime is strongly nival, with most of the annual runoff occurring in the spring and early summer. Winter flows are consistently low. The elevation ranges are very similar, 700–2370 m for Redfish Creek and 600–2340 m for Sitkum Creek; Redfish Creek has a slightly higher elevation distribution and therefore probably receives more snow. A potential important difference is that the Redfish Creek watershed has a history of forest development and is about 12% logged, while Sitkum Creek has had relatively little logging. However, the Sitkum Creek watershed contains an abandoned gold mine and mill, which was in production until 1948, with some exploration activity in the 1980s (BC Min. Energy & Mines, Minfile Mineral Inventory, http://minfile.gov.bc.ca/Summary.aspx?minfilno=082FNW127). The ruins of the mill burned in the fire, which led to some concern that heavy metal contamination could be a possibility. In addition to the stream gauging station near the mouth of Sitkum Creek, three small tributaries were identified, two within and one outside the burn, and V-notch weirs with water level recorders, pump samplers, and bedload traps were installed in August 2008.

Water quality sampling was undertaken in 2008 on Sitkum and Redfish creeks as part of a separate one-year project of the University of British Columbia (UBC), to investigate water quality indicators, at a province-wide scale, which may be relevant to mountain pine beetle infestation and other disturbances in community watersheds (Brown *et al.*, 2011). Samples were collected weekly during the spring freshet period, and monthly during the summer and autumn. The samples were analysed for nutrients, metals, and total organic carbon (TOC) at the soil science lab at UBC. Details of the equipment and methods used for analyses are given in Brown *et al.* (2011). In 2009 and 2010, a similar sampling programme was continued on Sitkum and Redfish creeks as part of

Fig. 2 Satellite image, showing the Sitkum Creek and Redfish Creek watersheds, and the locations of stream gauging stations. The extent of the 2007 Sitkum fire is outlined with a dashed line.

the present study. In 2009, samples were again analysed in the soil science laboratory at UBC. In 2010, this laboratory was not available, and samples were analysed for nutrients and TOC at a different laboratory at UBC, using similar methods and procedures. Unfortunately, inconsistencies in the 2010 data suggest that these results may not be reliable.

RESULTS

Soil erosion, turbidity and suspended sediment

A summary of sediment yields from the hillslope erosion plots is given in Table 1. These show a typical pattern of maximum erosion in the first year after the fire, with declining sediment yields each year thereafter as vegetation becomes established in the burned areas (Covert, 2010). The "0 years" data are for silt fence plots installed soon after the fire was out and sampled the same season; there was only one to two months after the fire until the first winter snow fell, and we had time to sample only two of the four fires.

These sediment yield results are generally lower than those reported elsewhere. In the western USA, a review by Moody & Martin (2009) showed that the highest post-wildfire sediment yields were at sites in Arizona and southern California, which also experience the highest short-term rainfall intensities. The lowest post-wildfire sediment yields and rainfall intensities were found in the "sub-Pacific" (or intermontane) region, which if extended north, would include the southern interior of BC. Other erosion plot results from the northernmost intermontane region in the USA (e.g. Robichaud *et al.*, 2009) are, on average, similar to the results from the present study. It should be noted that in this study, no rainstorms occurred in which short-term rainfall intensities (10 to 60 min) exceeded the 2-year return period.

At Sitkum Creek, the sediment yield results are anomalous, as the highest yields were measured in the second year. Visual observations of the plots indicated that a reason for this might be that much of the eroded sediment moved only a short distance downslope, and was then remobilized in subsequent rainfall or snowmelt events; it is likely that most erosion during the first year occurred in the upper part of the plots and did not reach the silt fence until the second year. The slope of the Sitkum plots was 31 to 38%, lower than the slopes of plots in the other fires, which ranged from 38 to 90%. Although there were only two plots installed in the Sitkum fire (there was no road access, and the sites required helicopter support to install and a long hike for each measurement), reconnaissance observations indicated that the results from these two plots were typical of erosion that occurred throughout the high-severity burn area.

In the three years following the 2007 fires, only one summer rainstorm occurred, on 19–21 August 2008, which produced enough runoff to cause a significant rise in the hydrographs of gauging stations in the region. At Sitkum Creek, there was 75 mm of rain in three days; the 24-h rainfall of 48 mm was about a 10-year event, but shorter-duration rainfall intensities were less than the 2-year return period. Although most of the soil erosion for the year probably occurred in this storm, the Sitkum Creek silt fences collected very little sediment (only 0.3 Mg/ha). However, the following year they collected much more (6 Mg/ha) even though there were no large rainstorms that summer. This supports the observation that most eroded sediment moved only a short distance downslope during each individual runoff event.

In the August 2008 rainstorms, on the two small tributary catchments of 48 and 64 ha draining the burned area, the automatic pump sampler collected samples every 2 hours, which with the continuous discharge data, enabled the suspended sediment yield for the storm to be calculated. The storm runoff was about 2 mm, less than 3% of the rainfall. The sediment yields for the two streams were 0.0005 and 0.0008 Mg/ha, at least three orders of magnitude lower than the surface erosion that probably occurred on severely burned sites during the storm. These results demonstrate that very little overland flow occurred in the burned area during this rainstorm, and that an insignificant amount, if any, of the eroded sediment reached the stream channels.

The stream gauges and pump samplers on Redfish and Sitkum creeks enable the calculation of approximate sediment yields for the two watersheds. The hydrographs, turbidity data, and annual sediment yields for 2008–2010 are shown in Fig. 3. In 2008, the suspended sediment yield for Sitkum Creek was 0.12 Mg/ha, which is almost three times the yield for Redfish Creek, although it is much lower than typical plot-scale sediment yields in the burned area. The turbidity data show that in 2008, Sitkum Creek had higher turbidity for much of the year, although turbidity on either creek rarely exceeded the common objective for domestic water sources of 5 NTU. In May 2008 (the exact date is unknown), a small landslide occurred on the mining road below the fire. It was probably caused by elevated groundwater levels below the burned area, and contributed some sediment (estimated as less than 100 m^3) to the creek. This event was probably responsible for the elevated turbidity in Sitkum Creek in May 2008. Another source of turbidity, especially later in the summer, may have been fine, burned, organic fragments from the burned area which were observed to accumulate in pools in the creek channel.

In 2009, the suspended sediment yield for Sitkum Creek was only slightly higher than for Redfish Creek, and except for isolated samples, turbidity on either creek rarely exceeded 5 NTU. In 2009, there were two significant sources of sediment in Sitkum Creek other than the fire. One of these was erosion of the road which was used for salvage logging in the summer of 2009, and contributed noticeable amounts of fine sediment to the creek as a result of maintenance and log hauling during wet weather. The other source was several large snow avalanches, which carried forest debris into the creek channel and created several log jams. In 2010, the suspended sediment yield for Sitkum Creek was double that of Redfish Creek. Considering the possible non-fire sediment sources, this may or may not be attributable to post-wildfire effects. It is possible, since late spring and summer flows in 2009 were very low, that sediment stored along the creek channel and in ephemeral headwaters channels in the first two years after the fire may have been remobilized by higher flows in 2010.

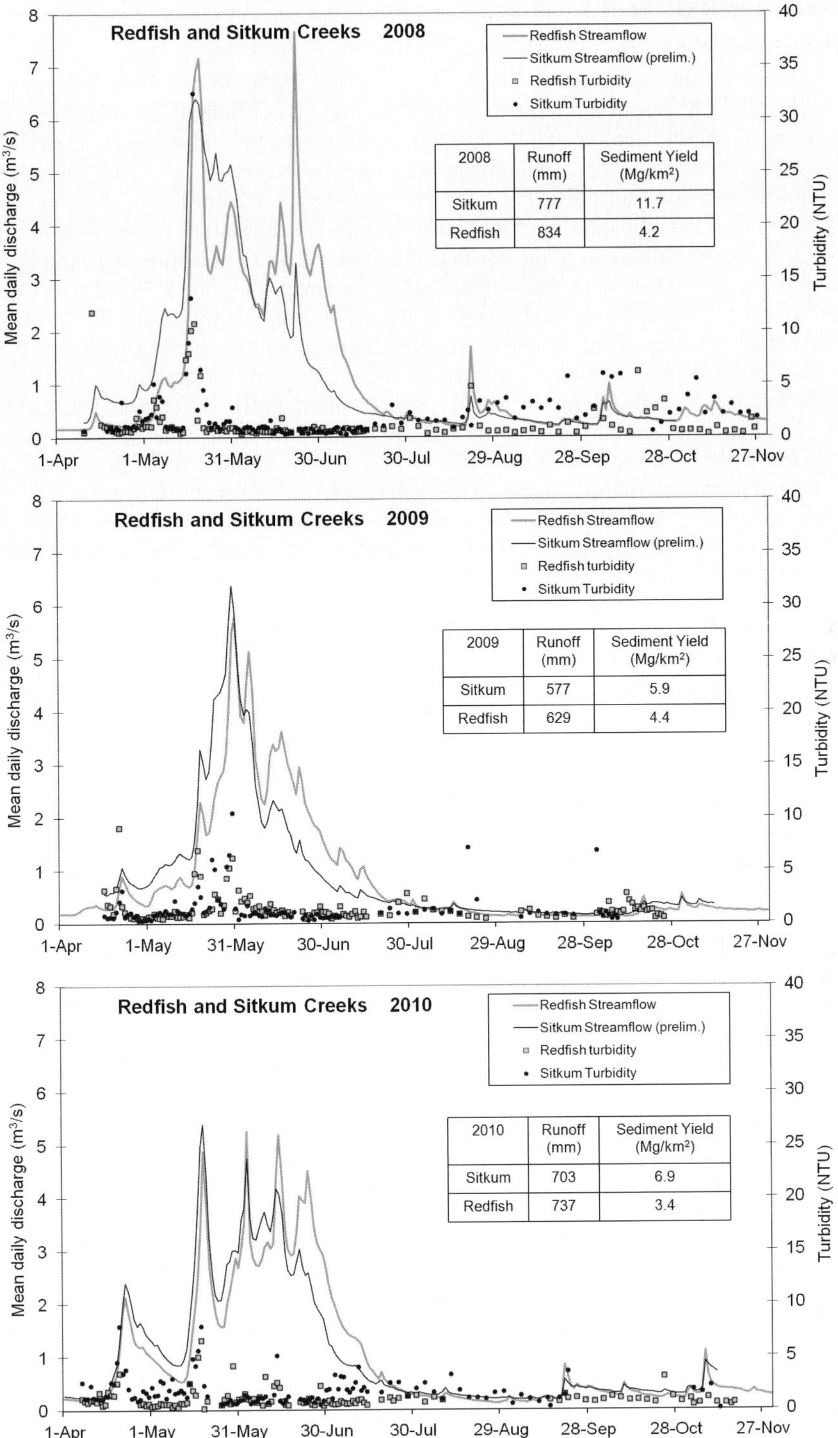

2008	Runoff (mm)	Sediment Yield (Mg/km²)
Sitkum	777	11.7
Redfish	834	4.2

2009	Runoff (mm)	Sediment Yield (Mg/km²)
Sitkum	577	5.9
Redfish	629	4.4

2010	Runoff (mm)	Sediment Yield (Mg/km²)
Sitkum	703	6.9
Redfish	737	3.4

Fig. 3 Discharge hydrographs for Redfish and Sitkum Creeks for the first to third years after the Sitkum fire, showing turbidity samples collected on the two creeks, and calculated runoff and suspended sediment yields.

CHEMICAL WATER QUALITY

Nitrogen (NO_x-N) was the only chemical water quality parameter which showed a significant difference between Sitkum and Redfish creeks. In 2008 and 2009, NO_x-N concentrations were about two to five times higher in Sitkum Creek, with the highest concentrations occurring in early spring and in late summer to autumn. These differences are consistent with other studies (Gluns & Toews, 1989; Bladon *et al.*, 2008; and others reviewed by Neary *et al.*, 2005) which have commonly reported increased nitrogen levels for several years after wildfire. In 2010, the NO_x-N data are less consistent, but still show slightly higher average concentrations for the year in Sitkum Creek. All NO_x-N concentrations were far less than the drinking water guideline of 10 mg/L.

Figure 4 shows the NO_x-N data, and Table 2 gives summary data for selected water quality parameters. No statistical tests were performed, as there is no replication or pre-fire data which could enable statistical comparison. Phosphorus (PO_4) showed no consistent difference between the two creeks, and most readings were close to, or below, the limit of detection. Total organic carbon (TOC) was inexplicably higher in Redfish Creek, although the difference is probably not meaningful. Calcium and sodium concentrations, pH, and electrical conductivity, were not noticeably different between the two creeks. Heavy metals showed little difference between the creeks (copper and iron were slightly higher in Redfish Creek); all concentrations were well below drinking water guidelines. Arsenic and lead (possible contaminants from the Sitkum Creek mine) were below detectable levels. The small differences between the creeks in parameters other than nitrogen and TOC can probably be attributed to minor differences in bedrock geology.

The results for parameters other than nitrogen differ from the other two studies from western Canada; both Gluns & Toews (1989) and Bladon *et al.* (2008) reported increased post-wildfire phosphorus concentration, and the former study also reported increased alkalinity and hardness. There is little information available on the effects of wildfire on releases of heavy metals (Neary *et al.*, 2005).

Table 2 Chemical water quality: mean, maximum, and standard deviation for selected parameters.

	Turbidity (NTU)	Conductivity (μS/cm, 25°C)	pH	NO_x-N (mg/L)	PO_4 (mg/L)	TOC (mg/L)	Ca (mg/L)	Na (mg/L)
Detection limit	0.2	–[1]	–	0.02	0.01		0.007	0.1
2008: n[2]	116	11						
Redfish: mean	1.12	23.6	6.81	0.059	0.015	3.39	3.32	0.63
maximum	10.8	–	–	0.155	0.043	7.69	6.15	1.09
st. dev.	1.70	10.3	0.26	0.048	0.011	1.65	1.36	0.26
Sitkum: mean	2.12	27.6	6.76	0.104	0.011	2.60	3.79	0.71
maximum	32.5	–	–	0.392	0.032	5.01	5.09	1.39
st. dev.	3.40	8.7	0.28	0.136	0.010	1.05	1.10	0.29
2009: n	106	8						
Redfish: mean	1.43	31.2	7.13	0.022	0.004	2.66	4.11	0.39
maximum	9.0	–	–	0.037	0.006	4.04	6.16	0.54
st. dev.	1.40	12.2	0.47	0.015	0.002	1.06	1.89	0.15
Sitkum: mean	1.34	32.8	7.00	0.115	0.005	2.15	3.99	0.42
maximum	10.4	–	–	0.270	0.014	3.26	5.25	0.58
st. dev.	1.60	8.9	0.13	0.099	0.005	0.70	1.17	0.13
2010: n	97	6						
Redfish: mean	1.08	25.2	6.82	0.110	0.049			
maximum	6.6	–	–	0.263	0.169			
st. dev.	0.98	7.8	0.66	0.099	0.061			
Sitkum: mean	1.87	24.3	6.65	0.167	0.036			
maximum	7.9	–	–	0.548	0.100			
st. dev.	1.40	10.1	0.83	0.203	0.034			

1. "–" indicates not applicable.
2. Sample size (n) each year is the same for all parameters except turbidity.
3. Blank cells indicate data are not available for that year.

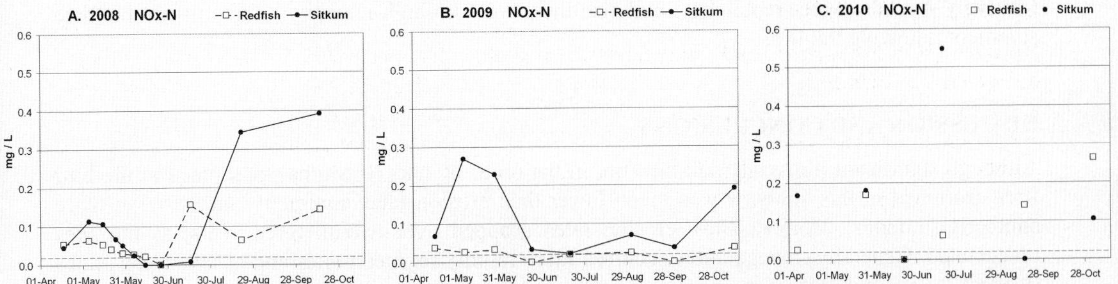

Fig. 4 Nitrate-nitrite (NO_x-N) samples collected on Sitkum and Redfish creeks, for three years after the 2007 Sitkum fire. Dashed line indicates the detection limit. (For 2010, no connecting line is shown, due to the lower frequency of sampling, and to one anomalous sample.)

ANECDOTAL ACCOUNTS OF POST-WILDFIRE IMPACTS ON WATER QUALITY

Sitkum Creek supplies water to a small community on its alluvial fan. The community water system consists of a water intake in the creek above the fan, and a sand-box filter to screen out debris and suspended sediment. Water from this filter then flows to an ultra-violet treatment facility. Regular water quality samples are collected at this facility, but not further upstream (D. Toews, Sitkum Creek Improvement District, pers. comm.). The only reported impact to the water supply was a greater than normal accumulation of suspended sediment and organic debris in the sand filter in 2008, which required more frequent back-flushing of the filter. In 2009, there was an unusual amount of debris in the filter, which necessitated replacement of the sand. As mentioned above, there were other sources of sediment in the water that year (road erosion and avalanches), so this sediment may not be entirely attributable to the fire.

There are two other cases in this region of large wildfires burning in community watersheds. The 2007 Kemp Creek fire burned about 20% of the community watershed supplying the village of Kaslo. The 2003 Kutetl fire burned about 10% of the drainage area of Five-Mile Creek, the main community watershed supplying the city of Nelson. In neither of these cases were there reports of increases in turbidity, other water quality effects, or physical impacts to the water intakes, affecting the municipal water supplies (G. Walker, public works foreman, Village of Kaslo, pers. comm.; G. Bogaard, utilities supervisor, City of Nelson, pers. comm.). These observations are in contrast with reports of large sediment yields following wildfire at some locations in the USA, especially Colorado and southern California (as reviewed by Moody & Martin, 2009). However, they are consistent with the general lack of reports of significant post-wildfire turbidity increases or other water quality impacts from water utilities in BC.

The summer following the 2003 Kuskonook fire near Creston, water users on Kuskonook Creek (a domestic watershed) noticed large quantities of a black substance in their water intakes. A month later, when the author and others investigated the large debris flow on Kuskonook Creek (Jordan & Covert, 2009) we observed areas where the wettable layer of burned surface soil had been eroded away, above a water repellent layer which was 1–2 cm below the surface. This fire exhibited strong and extensive water repellency, as did several other 2003 fires which burned under extreme drought conditions. The Kuskonook Creek example is the only case in this region known to the author where a significant water quality impact was reported following the 2003 or subsequent fires (and it became moot later that summer when the debris flow destroyed all water intakes on the creek). However, there is no systematic reporting or data collection of water quality on domestic watersheds, so it is possible that unreported cases of poor water quality may have occurred.

Our observations at Kuskonook Creek suggest that strong water repellency may be an important factor in the likelihood that burned surface soil and ash will be transported to stream channels during high-intensity rainfall events. The 2007 fires in this region, in contrast, had relatively sporadic water repellency, and also there were no high-intensity rainstorms in the three summers following the fires.

DISCUSSION AND CONCLUSIONS

Although significant plot-scale soil erosion, in the order of 1 to 10 Mg/ha, took place in the four fires examined in this study, it was much lower than erosion rates commonly measured at more southerly latitudes. At most sites, erosion rates dropped over two to three years, as the sites revegetated. Generally, revegetation occurred more rapidly on moist sites and at lower elevations (Covert, 2010), and erosion rates were accordingly lower on these sites. Since 2007, high-intensity short-duration rainstorms have been lacking; if such rainstorms had occurred within one or two years of the fires, it is likely that erosion rates would have been higher. After the 2003 wildfires in southern British Columbia, in the first year after the fire there were unusually abundant high-intensity rainstorms in the region, and as a result there were several severe landslide and flooding incidents.

In the 2007 Sitkum Creek fire, sediment yield measurements were made at two watershed scales, small (<100 ha) tributaries, and the whole catchment (27 km^2), and comparisons were made with the similar Redfish Creek watershed. For three years after the fire, both physical and chemical water quality impacts were minimal. At both scales, increases in turbidity and suspended sediment yield occurred in the first year after the fire, but these were slight, and comparable to increased erosion and sedimentation that typically is caused in this region by other disturbances such as logging and road building (Jordan, 2006).

One reason for the relatively slight impacts (compared with other regions at lower latitudes) of wildfire on erosion and sedimentation in the southern interior of British Columbia is the strongly nival runoff regime (Eaton *et al.*, 2010). The annual streamflow hydrograph for all but very small or low-elevation drainages is invariably dominated by snowmelt, and even with increased rainstorm runoff following wildfire, it is rare for discharge during summer events to approach the spring snowmelt peak. Therefore, entrainment of sediment from gully down-cutting and bank erosion is unlikely during summer rainstorm events. Also, at higher elevations, most precipitation falls as snow, and the snow-free season is only four to six months, so there is less time each year for rainfall events to occur than in regions with relatively little snow cover.

Another reason for the relatively low sediment yields observed in this study is that, across most of the landscape, the connectivity between hillslopes and stream channels is low. This may be a result of the glacial history of the region, which has resulted in many U-shaped or widened valleys and under-fit streams, in contrast to unglaciated, fluvially-dissected, landscapes. Qualitative observations in burned areas during this study suggested that in most locations, eroded sediment moved only a short distance downslope and was deposited behind obstructions, in stony areas, on gentler slopes in local valley bottoms, or behind rapidly-growing riparian vegetation.

These general conclusions about low sediment yields following fire must be qualified with the observation that, in susceptible terrain, the likelihood of debris flows and other landslides can be substantially increased by severe wildfire (Cannon & Gartner, 2005; Shakesby & Doerr, 2006; Jordan & Covert, 2010; Jordan, 2012). In this region, the increased landslide hazard is largely due to higher snow accumulation and snowmelt rates following wildfire, and consequent higher groundwater levels. If a debris flow occurs, inevitably there will be a significant impact on water quality, although this may be short-lived.

Chemical water quality impacts on Sitkum Creek, relative to Redfish Creek, were limited to an increase in nitrogen (as NO_x-N), consistent with measurements that have been frequently reported following wildfire elsewhere. Differences in other solutes were minimal or nonexistent, compared with a nearby unburned reference watershed. These results are consistent with conclusions reported elsewhere, that changes in post-wildfire chemical water quality are seldom a concern for potable water supplies; however, potential increases in turbidity and sediment yield should be considered in managing community water systems after wildfire.

Acknowledgements Ashley Covert and Amy O'Neill participated in the field work and data analysis on this project. Sandra Brown, Faculty of Land and Food Systems, University of British Columbia, designed the sampling programme for chemical water quality and arranged for sample analysis at UBC. Funding for three years from 2007 to 2010 was provided by a BC Forest Investment Account grant. This paper benefited from a thorough review by Adrian Collins.

REFERENCES

Beschta, R. L. (1990) Effects of fire on water quantity and quality. In: *Natural and Prescribed Fire in Pacific Northwest Forests* (ed. by J. D. Walstad, S. R. Radosevich & D. V.Sandberg), 219–232. Oregon State University Press, Corvallis, Oregon, USA.

Bladon, K. D., Silins, U., Wagner, M. J., Stone, M., Emelko, M. B., Mendoza, C. A., Devito, K. J. & Boon, S. (2008) Wildfire impacts on nitrogen concentration and production from headwater streams in southern Alberta's Rocky Mountains. *Can. J. Forest Res.* 38, 2359–2371.

British Columbia Ministry of Environment, Environmental Protection Division (undated) *Water Quality Guidelines (Criteria) Reports*. http://www.env.gov.bc.ca/wat/wq/wq_guidelines.html (accessed 20 Jan. 2012).

Brown, S., Lavkulich, L. M. & Schreier, H. (2011) Developing indicators for regional water quality assessment: an example from British Columbia community watersheds. *Can. Water Resour. J.* 36, 271–284.

Cannon, S. H. & Gartner, J. E. (2005) Wildfire-related debris flow from a hazards perspective. In: *Debris-flow Hazards and Related Phenomena* (ed. by M. Jakob & O. Hungr), 363–384. Springer-Praxis, Berlin.

Covert, S. A. (2010) The effects of straw mulching on post-wildfire vegetation recovery in southeastern British Columbia. *BC J. Ecosystems and Management* 11(3), 1–12.

Covert, S. A. & Jordan, P. (2009) A portable rainfall simulator: techniques for understanding the effects of rainfall on soil erodibility. *Streamline Watershed Management Bull.* 13(1), 5–9.

Doerr, S. H., Shakesby, R. A., Blake, W. H., Chafer, C. J., Humphreys, G. S. & Wallbrink, P. J. (2006) Effects of differing wildfire severities on soil wettability and implications for hydrological response. *J. Hydrol.* 319(1-4), 295–311.

Eaton, B. C., Moore, R. D. & Giles, T. R. (2010) Forest fire, bank strength and channel instability: the 'unusual' response of Fishtrap Creek, British Columbia. *Earth Surf. Processes Landf.* 35, 1167–1183.

Filmon, G., Leitch, D. & Sproul, J. (2004) Firestorm 2003 Provincial Review. Report to the Premier, Province of British Columbia, Victoria, BC.

Gluns, D. R. & Toews, D. A. A. (1989) Effect of a major wildfire on water quality in southeastern British Columbia. In: *Headwaters Hydrology* (ed. by W. W. Woessner & D. F. Potts), 487–499. Am. Water Resour. Assoc., Bethesda, Maryland, USA.

Jordan, P. (2006) The use of sediment budget concepts to assess the impact on watersheds of forestry operations in the Southern Interior of British Columbia. *Geomorphol.* 79, 27–44.

Jordan, P. (2012) Post-wildfire landslides in southern British Columbia. In: Proc. 11th International & 2nd North American Symposium on Landslides (3–8 June 2012, Banff, Alberta) (in press).

Jordan, P. & Covert, S. A. (2009) Debris flows and floods following the 2003 wildfires in southern British Columbia. *Environ. Eng. Geosci.* 15, 217–234.

Moody, J. A. & Martin, D. A. (2009) Synthesis of sediment yields after wildland fire in different rainfall regimes in the western United States. *Int. J. Wildland Fire* 18, 96–115.

Neary, D. G., Landsberg, J. D., Tiedemann, A. R. & Ffolliott, P. F. (2005) Water Quality. In: *Wildland Fire in Ecosystems: Effects of Fire on Soils and Water* (ed. by D. G. Neary, K. C. Ryan & L. F. DeBano), Chapter 6, 119–134. US Dept. Agric., Forest Service, Rocky Mountain Res. Stn., RMRS-GTR-42, vol. 4. Ogden, Utah, USA.

Pike, R. G., Feller, M. C., Stednick, J. D., Rieberger, K. J. & Carver, M. (2010) Water quality and forest management. In: *Compendium of Forest Hydrology and Geomorphology in British Columbia* (ed. by R. G. Pike, T. E. Redding, R. D. Moore, R. D. Winker & K. D. Bladon) Chapter 12, 401–440. Land Management Handbook 66, BC Ministry of Forests and Range, Victoria, BC and FORREX Forum for Research and Extension in Natural Resources, Kamloops, BC, Canada

Robichaud, P. R., Beyers, J. L. & Neary, D. G. (2000) Evaluating the effectiveness of postfire rehabilitation treatments. *US Dept. Agric., Forest Service, Rocky Mountain Res. Stn, RMRS-GTR-63*. Fort Collins, Colorado, USA.

Robichaud, P. R., Wagenbrenner, J. W., Brown, R. E. & Spigel, K. M. (2009) Three years of hillslope sediment yields following the Valley Complex fires, western Montana. *US Dept. Agric., Forest Service, Rocky Mountain Res. Stn, RMRS-RP-77*. Fort Collins, Colorado, USA.

Robichaud, P. R. & Brown, R. E. (2002) Silt fences: an economical technique for measuring hillslope soil erosion. *US Dept. Agric., Forest Service, Rocky Mountain Res. Stn, RMRS-GTR*-94. Fort Collins, Colorado, USA.

Shakesby, R. A. & Doerr, S. H. (2006) Wildfire as a hydrological and geomorphological agent. *Earth-Sci. Rev.* 74, 269–307.

Silins, U., Bladon, K. D., Anderson, A., Diiwu, J., Emelko, M. B., Stone, M. & Boon, S. (2009) Alberta's southern Rockies watershed project – how wildfire and salvage logging affect water quality and aquatic ecology. *Streamline Watershed Management Bull.* 12(2), 1–7.

Smith, H. G., Sheridan, G. J., Lane, P. N. J., Nyman P. & Haydon, S. (2011) Wildfire effects on water quality in forest catchments: a review with implications for water supply. *J. Hydrol.* 396, 170–192.

Spittlehouse, D. (2006) ClimateBC: Your access to interpolated climate data for BC. *Streamline Watershed Management Bull.* 9(2), 16–21.

The effect of in-stream wood structures on fine sediment storage in headwater streams of the Canadian Rocky Mountains

KATHLEEN LITTLE[1], MIKE STONE[2] & ULDIS SILINS[3]

1 *School of Planning, University of Waterloo, Waterloo, Ontario N2L 3G1, Canada*
k2little@uwaterloo.ca

2 *Dept. of Geography and Environmental Management, University of Waterloo, Waterloo, Ontario N2L 3G1, Canada*

3 *Dept. of Renewable Resources, University of Alberta, Edmonton, Alberta T6G 2H1, Canada*

Abstract The recruitment of wood from channel margins to headwater streams can profoundly influence stream hydrology, morphology, ecology and sediment transport dynamics (erosion and deposition). This study examines the effect of in-stream wood deposits on pool formation and fine sediment storage in two headwater streams with contrasting land disturbance on the eastern slopes of the Canadian Rocky Mountains. Six representative study reaches, three from each disturbance type (burned/salvage logged *vs* recreation/grazing) were selected to examine pool spacing, pool type and fine sediment storage. A total of 94 and 66 in-stream wood structures were found in the Lyons East (LE) and Corolla Creek (CC) study areas. The frequency of in-stream wood structures was 1.5/100 m (LE) and 1.48/100 m (CC). Overall, the volume of fine sediment stored in pools was greater in burned-salvage logged reaches (Lyons East) compared to unburned (reference) reaches of Corolla Creek. The increased volume of fine sediment is related to increased sediment availability due to mass wasting processes and overland flow caused by wildfire.

Key words in-stream wood; dams; jams; fine sediment; gravel bed rivers; wildfire; watershed disturbance; Canadian Rocky Mountains

INTRODUCTION

Forested watersheds in western North America provide a wide range of ecosystem services and are critical source regions for the drinking water supply of more than 180 million people (Hauer *et al.*, 2007; Emelko *et al.*, 2010). The effects of landscape disturbance (recreation, logging, wildfire, agriculture) on the morphology and ecological integrity of streams draining forested landscapes (Fetherston *et al.*, 1995; Gurnell *et al.*, 1995; Whol & Jaeger, 2009) are well documented. Wildfire can accelerate the rates and magnitudes of terrestrial erosion and sediment loading into fire affected watersheds (DeBano *et al.*, 1998; Silins *et al.*, 2009), which along with increased wood recruitment from burned channel margins can lead to in-stream wood structures that often produce secondary effects of wildfire on stream morphology, ecology and sediment transport dynamics (Beschta & Platts, 1986; Nakamura *et al.*, 2000). The input of wood to streams also changes channel bed roughness, which can dissipate stream energy and promote sediment deposition upstream of the obstruction (Bilby & Ward, 1989; Trotter, 1990; Curran & Wohl, 2003; Webb & Erskine, 2003; Gurnell *et al.*, 2005).

Various forms of in-stream wood structures, such as dams and jams, influence the amount and type of sediment stored in the host channel (Jackson & Strum, 2002; May & Gresswell, 2003). Fine sediment (clay, silt, fine sand) is typically stored in the void spaces of natural gravel bed rivers (Collins & Walling, 2007). When the supply of sediment exceeds the storage capacity of the gravel bed, continued fine sediment deposition causes surficial layers or drapes to form in pools during conditions of low flow (Lisle & Hilton, 1999). According to the work of Lisle & Hilton (1992), the fraction of pool volume filled with fine sediment can be used as an index of the supply of mobile sediment. They define this fraction (V^*) as the ratio of fine sediment to pool water volume plus fine-sediment volume. V^* has been used as an index of the supply of mobile sediment in the stream channel (Lisle & Hilton, 1992) as well as an indicator of habitat suitability for aquatic organisms (Hilton & Lisle, 1993). Although this approach has been used to study the effects of wildfire on fine sediment storage in Wyoming streams (Zelt & Wohl, 2004), no comparable studies have been conducted in wildfire impacted landscapes in the Canadian Rockies. Such information is required to evaluate and compare the longer term, secondary effects of wildfire on fine sediment storage in gravel bed streams which, in turn, has implications for water quality and habitat suitability and sediment sources. The objective of this study was therefore to examine the effect of in-stream wood structures on fine sediment storage in two catchments of varying land disturbance types (burned *vs* unburned reference) that are located in the headwater regions of the eastern slopes of the Canadian Rocky Mountains.

METHODS

The study was conducted in two headwater catchments of the Oldman River Basin (Lyons East 49°34′08.5″N, 114°27′19.4″W, and Corolla Creek 49°25′30.6″N, 114°23′26.8″W) located in southwest Alberta. The two catchments are located in close proximity to one another (~10 km) and are situated in similar hydro-climatic and physiographic settings just east of the continental divide along the Flathead range. Mean annual precipitation is approximately 850 mm/year for both catchments. The drainage area and perimeter of the study watersheds are comparable but Lyons East has a larger average channel and catchment slope (Table 1). The studied streams in both catchments are classified as third order channels. Lyons East (LE) was initially disturbed by wildfire in 2003, then subsequently post-fire salvage logged (82% burned, 24% of which was salvage logged; no logging occurred within 30 m of the stream). Corolla Creek (CC) is an unburned reference watershed. Soils in the study catchments include thin gray luvisol and brunisol soils, which influence the type of vegetation in this mountain eco-region (Bladon *et al*., 2008; Oldman Watershed Council, 2010).

Table 1 Physical characteristics of the study catchments.

Study basin	Drainage area (km^2)	Perimeter (km)	Mean elevation (m) (Range)	Catchment slope (%)	Channel slope
Burned	13.06	17.2	1683 (1441–2028)	31.4	0.04
Reference	13.15	20.3	1568 (1429-1963)	11.4	0.028

An initial survey was conducted using GPS to identify the location, type and frequency of all in-stream wood structures in the study basins. Three representative in-stream wood structures (Fig. 1) were identified in each study catchment. Detailed surveys were conducted to classify wood structure type and morphological properties. Pool type was classified according to the formation mechanism (self-forming or forced) and channel processes (scouring or damming) using the methods of Hawkins *et al.* (1993) and Montgomery *et al.* (1995). Pool dimensions (length, width and depth) were determined using the methods outlined by Hilton & Lisle (1993). Pool volume was calculated by measuring water depth at a minimum of 10 locations across each pool, and the number of transects surveyed varied among pools as a function of pool length. Typically, transects were located 1 m apart for longer pools and 0.5 m apart for shorter pools.

The relative volume of fine sediment in pools (V*) was described as the fraction of scoured pool volume occupied by fine sediment (Lisle & Hilton, 1992). Measurements of water depth, sediment surface depth and fine sediment depth were made at a minimum of 10 intervals across each pool transect using a steel metre stick with an accuracy of 0.1 cm. An abrupt change in resistance to the rod as it passed from sand or fine gravel to packed coarse gravel and cobbles indicated the base of the residual pool. The weighted average (V_w*) of all pools in each study reach was calculated using the following equation.

$$V_w^* = \sum(\text{residual fines volume})/\sum(\text{scoured residual pool volume}) \quad (1)$$

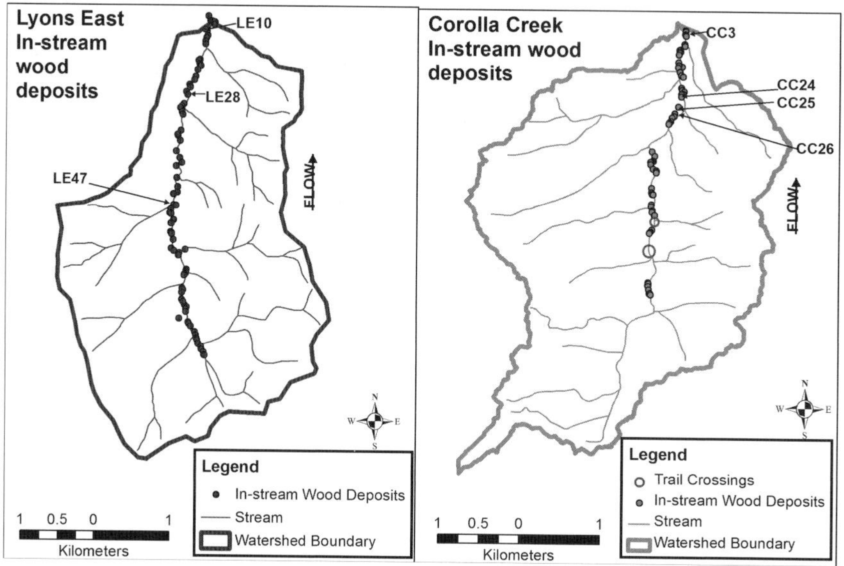

Fig. 1 Location of all in-stream wood structures and detailed study sites in Lyons East and Corolla study catchments.

RESULTS AND DISCUSSION

Wood structures and pool morphology

The location of all in-stream wood structures and detailed study sites (LE10, LE28, LE47) (CC3, CC24, CC26) are shown in Fig. 1. A total of 94 and 66 in-stream wood deposits that included jams, partial jams and dams were identified (Table 2). Although there was a difference in the total number of in-stream wood structures, the frequency of in-stream wood deposits was estimated at 1.5/100 m (LE) and 1.48/100 m (CC). This estimate is comparable to the work of Zelt & Wohl (2004). They reported frequencies of 1.6 and 1.5 debris jams/100 m in burned and reference catchments, respectively. For comparison, Wohl & Jaeger (2009) reported fewer than 2 jams/100 m in a study conducted on the Colorado Front Range.

In-stream wood can have a profound effect on channel morphology and pool formation (Kreutzweiser *et al.*, 2005). Previous studies describe a relationship between wood loading and pool spacing in plane bed, pool-riffle and forced pool-riffle channels (Montgomery *et al.*, 1995; Abbe & Montgomery, 1996). In a typical free-formed pool-riffle stream, pools are spaced approximately five to seven channel widths apart (Leopold *et al.*, 1964) compared to one to four channel widths typical of step-pool channel forms (Montgomery & Buffington, 1997). In the current study, pool spacing ranged from 1.53 to 2.31 channel widths (Table 3) which is less than expected for the pool-riffle streams we studied, and at the lower end for streams with step-pool morphology.

Table 2 Number and frequency of in-stream wood structures.

Wood structure	Burned count	Burned frequency (/100 m)	Reference count	Reference frequency (/100 m)
Single	10	0.16	9	0.2
Partial jam	18	0.29	13	0.29
Jam	40	0.64	31	0.7
Dam	26	0.41	13	0.29
Total	94	1.5	66	1.48

Table 3 Summary of reach-average channel characteristics.

Channel characteristic	Symbol	LE10	LE28	LE47	CC3	CC24	CC26
Bankfull width (m)	W_{bf}	13.22	10.23	12.99	12.67	12.95	14.73
Gradient (m/m)	S_w	0.03	0.02	0.02	0.02	0.02	0.03
Forced Pools (%)	F%	67	60	67	20	33	100
Forced pools – wood-affected (%)	F_{wa}%	50	20	33	20	33	33
Pool length (m)	PL_m	37	25	13	43	11	17
Pool length (channel widths)	PL	2.8	2.44	1	3.39	0.85	1.15
Pool spacing (m)	SP_m	20	20	30	21	29	31
Pool spacing (channel widths)	SP	1.53	1.97	2.31	1.63	2.24	2.1
% reach length in pools	P%	30	24	14	42	12	19
Mean residual pool depth (cm)	D_r	10.6	7.3	13.2	12.5	14	18.9
Mean residual volume per pool (m^3)	V_r	1.21	1.27	1.64	2.95	3.34	3.99

Table 4 Residual pool and fine sediment volume.

Site	Pool #	Location	Pool type	Residual pool volume (m^3)	Residual sediment volume (m^3)	V*
LE10	1	US	Lateral scour	1.094	0.002	0.001
	2	US	Mid-channel	1.291	0.004	0.003
	3	At wood	Plunge	0.49	None	
	4	At wood	Deflector	1.699	0.016	0.01
	5	At wood	Underflow	1.361	0.006	0.004
	6	DS	Plunge	1.294	0.006	0.005
LE28	1	US	Plunge	0.472	None	
	2	US	Mid-channel	0.754	None	
	3	US	Lateral scour	0.466	None	
	4	At wood	Deflector	4.43	0.379	0.085
	5	DS	Plunge	0.239	None	
LE47	1	US	Lateral scour	1.241	0.042	0.033
	2	At wood	Underflow	3.1	0.049	0.016
	3	DS	Lateral scour	0.588	0.002	0.004
CC3	1	US	Trench	1.413	None	
	2	US	Trench	3.542	None	
	3	At wood	Dam	1.844	None	
	4	DS	Mid-channel	3.171	None	
	5	DS	Trench	4.785	None	
CC24	1	At wood	Underflow	3.729	0.017	0.004
	2	Between	Lateral scour	1.229	None	
	3	DS	Mid-channel	5.063	0.154	0.03
CC26	1	US	Deflector	0.014	None	
	2	At wood	Deflector	6.213	0.011	0.002
	3	DS	Trench	5.73	None	

Wood structure played a key role in pool characteristics in each of the six study reaches (Table 4). In half of the study reaches (LE28, LE47 and CC26), the highest pool volumes were found at in-stream wood structures. The wood-affected pool volume was approximately 50 to 70% of the total pool volume compared to 12.5 to 52% in Corolla Creek.

Fine sediment storage

The presence of wood structures was also closely associated with residual volume of fine sediment (V*) measured in pools (Table 4). With the exception of site CC3, fine sediment deposits were consistently observed at pools created by in-stream wood structures. The largest residual fine sediment deposit was measured at the inlet of pool 3 (site CC24) which had experienced considerable bank failure in the reach immediately upstream. The second largest deposits of residual fine sediment were located in the middle and upper reaches of Lyons East Creek (sites LE28 and LE47). Residual deposits of fine sediment were also observed at sites CC3 and CC26. With the exception of a plunge pool (LE10–pool 3), the highest volume of fine sediment storage occurred at wood-affected pools in Lyons East Creek. Accordingly, the majority (5 out of 8) of wood-affected pools were found to store more fine sediment than in non-wood-affected pools.

The weighted average relative volume of fine sediment in pools (V^*_w) was also affected by wildfire (Table 5). Zelt & Wohl (2004) reported an average residual pool fine sediment volume (V^*_w) of 0.17 and 0.23 for reference and burned streams, respectively. In the present study, V^*_w ranged from 0.005 to 0.056 in burned, and 0 to 0.017 in unburned streams. While V^*_w was highly variable between in-stream wood structures, it was more pronounced in Lyons East Creek indicating a higher presence of stored sediment in the stream affected by wildfire disturbance.

In the current study, the average sediment depth in pools ranged from <0.1 to 0.7cm and from <0.1 to 0.3cm in Lyons East and Corolla Creeks, respectively. The presence of more sediment is likely related to the effect of land disturbance by wildfire on sediment source, availability and conveyance processes. Based on field observations, sediment supply to the channel was more pronounced due to vegetation loss and its impact on overland flow hillslope/bank failure. Previous research on in-stream wood and associated fine-grained sediment storage reported thicker but less frequent deposits within disturbed watersheds (Zelt & Wohl, 2004). Approximately 64% of the pools burned streams contained deposits of fine sediment compared to 36% of the pools in reference streams.

Table 5 Weighted average of V* for selected study reaches.

Disturbance	Site	V^*_w	n
Burned	LE10	0.005	6
	LE28	0.056	5
	LE47	0.019	3
Reference	CC3	0	5
	CC24	0.017	3
	CC26	0.001	3

CONCLUSION

The volume of fine sediment deposits was evaluated in pools of gravel bed streams draining two study catchments with contrasting landscape disturbance. The sediment deposits were generally thicker and more prevalent in pools of the wildfire impacted stream compared to the reference stream. The lack of sediment in Corolla Creek was likely also related to the greater presence of bedrock in the channel, which limited the sediment supply. The results suggest that pools located at in-stream wood deposits represent an important sink for fine sediment in highly disturbed headwater streams on the eastern slopes of the Canadian Rocky Mountains, which may impact local and downstream water quality and aquatic habitats.

Acknowledgements Funding for the research was provided by AWRI/AIEES, and NSERC grants to M. Stone. Funding was also provided by an NSERC grant to K. Little. Support for this research by Alberta Sustainable Resource Development (forest management division) is also gratefully acknowledged.

REFERENCES

Abbe, T. & Montgomery, D. (1996) Large woody debris jams, channel hydraulics and habitat formation in large rivers. *Regulated Rivers: Research & Management,* 12(2-3), 201–221.

Beschta, R. L. & Platts, W. S. (1986) Morphological features of small streams: Significance and function. *JAWRA, J. Am. Water Resour. Assoc.* 22(3), 369–379.

Bilby, R. E. & Ward, J. W. (1989) Changes in characteristics and function of woody debris with increasing size of streams in western Washington. *Trans. American Fisheries Society* 118, 368–378.

Bladon, K. D., Silins, U., Wagner, M. J., Stone, M., Emelko, M. B., Mendoza, C. A., Devito, K. J. & Boon, S. (2008) Wildfire impacts on nitrogen concentration and production from headwater streams in southern Alberta's Rocky Mountains. *Can. J. Forest Res.* 38(9), 2359–2371.

Collins, A. L. & Walling, D. E. (2007) Fine-grained bed sediment storage within the main channel systems of the Frome and Piddle catchments, Dorset, UK. *Hydrol. Processes* 21, 1448–1459.

Curran, J. H. & Wohl, E. E. (2003) Large woody debris and flow resistance in step-pool channels, Cascade Range, Washington. *Geomorphology* 51(1-3), 141–157.

DeBano, L. F., Neary, D. G. & Ffolliott, P. F. (1998) *Fire's Effects on Ecosystems.* John Wiley and Sons, New York, NY.

Emelko, M., Silins, U., Bladon, K. & Stone, M. (2010) Implications of land disturbance on drinking water treatability in a changing climate: demonstrating the need for "source water supply and protection" strategies. *Water Research* 45(2), 461–472.

Fetherston, K. L., Naiman, R. J. & Bilby, R. E. (1995) Large woody debris, physical process, and riparian forest development in montane river networks of the Pacific Northwest. *Geomorphology* 13, 133–144.

Gurnell, A. M., Gregory, K. J. & Petts, G. E. (1995) The role of coarse woody debris in forest aquatic habitats: Implications for management. *Aquatic Conservation: Marine and Freshwater Ecosystems* 5, 143–166.

Gurnell, A. M., Tockner, K., Edwards, P. J. & Petts, G. (2005) Effects of deposited wood on biocomplexity of river corridors. *Frontiers in Ecology and the Environment* 3, 377–382.

Hauer, F. R., Stanford, J. A. & Lorang, M. S. (2007) Pattern and process in northern Rocky Mountain headwaters: Ecological linkages in the headwaters of the Crown of the Continent1. *JAWRA J. Am. Water Resour. Assoc.* 43(1), 104–117.

Hawkins, C. P., Kershner, J. L., Bisson, P. A., Bryant, M. D., Decker, L. M., Gregory, S. V., McCullough, D. A., Overton, C. K., Reeves, G. H., Steedman, R. J. & Young, M. K. (1993) A hierarchical approach to classifying stream habitat features. *Fisheries* 18(6), 3–10.

Hilton, S. & Lisle, T. E. (1993) *Measuring the fraction of pool volume filled with fine sediment* (Report no. PSW-RN-414). Albany, CA: Pacific Southwest Forest and Range Experiment Station, Forest Service, US Department of Agriculture; 11 p.

Jackson, C. R. & Strum, C. A. (2002) Woody debris and channel morphology in first- and second-order forested channels in Washington's coast ranges. *Water Resour. Res.* 38(9), 1–14.

Keller, E. A. & Swanson, F. J. (1979) Effects of large organic material on channel form and fluvial processes. *Earth Surface Processes and Landforms* 4, 351–380.

Kreutzweiser, D. P., Good, K. P. & Sutton, T. M. (2005) Large woody debris characteristics and contributions to pool formation in forest streams of the boreal shield. *Can. J. Forest Res.* 35, 1213–1223.

Leopold, L. B., Wolman, M. G. & Miller, J. P. (1964) *Fluvial Processes in Geomorphology.* W. H. Freeman and Company, San Francisco.

Lisle, T. E. & Hilton, S. (1992) The volume of fine sediment in pools: an index of sediment supply in gravel-bed streams. *Water Resources Bull.* 28(2), 371–383.

Lisle, T. E. & Hilton, S. (1999) Fine bed material in pools of natural gravel bed channels. *Water Resour. Res.* 35(4), 1291–1304.

May, C. L. & Gresswell, R. E. (2003) Processes and rates of sediment and wood accumulation in headwater streams of the Oregon Coast Range, USA. *Earth Surface Processes and Landforms* 28(4), 409–424.

Montgomery, D. R. & Buffington, J. M. (1997) Channel-reach morphology in mountain drainage basins. *Geol. Soc. Am. Bull.* 109(5), 596–611.

Montgomery, D. R., Buffington, J. M., Smith, R. D., Schmidt, K. M. & Pess, G. (1995) Pool spacing in forest channels. *Water Resour. Res.* 31(4), 1097–1105.

Nakamura, F., Swanson, F. J. & Wondzell, S. M. (2000) Disturbance regimes of stream and riparian systems – A disturbance-cascade perspective. *Hydrol. Processes* 14(16-17), 2849–2860.

Oldman Watershed Council. (2010) *Oldman River State of the Watershed Report 2010.* Oldman Watershed Council, Lethbridge, Alberta: 284 pp.

Silins, U., Stone, M., Emelko, M. B. & Bladon, K. D. (2009) Sediment production following sever wildfire and post-fire salvage logging in the Rocky Mountain headwaters of the Oldman River Basin, Alberta. *Catena*, 79(3), 189–197.

Trotter, E. H. (1990) Woody debris, forest–stream succession, and catchment geomorphology. *J. North Am. Benthological Society* 9(2), 141–156.

Webb, A. A. & Erskine, W. D. (2003) Distribution, recruitment, and geomorphic significance of large woody debris in an alluvial forest stream: Tonghi Creek, southeastern Australia. *Geomorphology* 51, 109–126.

Wohl, E. & Jaeger, K. (2009) A conceptual model for the longitudinal distribution of wood in mountain streams. *Earth Surface Processes and Landforms* 34, 329–344.

Zelt, R. B. & Wohl, E. E. (2004) Channel and woody debris characteristics in adjacent burned and unburned watersheds a decade after wildfire, Park County, Wyoming. *Geomorphology* 57(3-4), 217–233.

Hillslope erosion and post-fire sediment trapping at Mount Bold, South Australia

ROWENA MORRIS[1,2], DEIRDRE DRAGOVICH[3] & BERTRAM OSTENDORF[1]

1 *Earth and Environmental Science, PMB 1, Glen Osmond, University of Adelaide, South Australia 5064, Australia*
rowena.morris@adelaide.edu.au

2 *Bushfire Cooperative Research Centre, Level 5, 340 Albert Street, East Melbourne, Victoria 3002, Australia*

3 *School of Geosciences, Madsen Building (F09), University of Sydney, New South Wales 2006, Australia*

Abstract Successful placement of sediment traps requires an understanding of how hillslope morphology influences erosion. Following the 2007 Mount Bold wildfire, in South Australia, a 1 in 5 year rainfall event resulted in the failure of many sediment traps due to substantial sediment movement within the reservoir reserve. This study assesses how hillslope morphology can influence post-fire surface erosion and the subsequent appropriate placement of sediment traps. Erosion pins and sediment traps were used at five different sites to measure hillslope surface change and trapped sediment volumes. Terrestrial laser scanning was used to model surface change where slope gradients are 1:2 or greater. Surface change was assessed in relation to slope gradient, slope length, cross-slope curvature, hillslope position and fire severity. The results suggested a threshold for substantial increased sediment yield at slope gradients of 1:2. The findings also suggested that concave cross-slope curvatures were associated with significantly larger amounts of sediment movement.

Key words water reservoir; sediment trap; erosion pins; terrestrial laser scanning; slope gradient; cross-slope plan curvature, South Australia

INTRODUCTION

Wildfires influence soil surface processes resulting in the potential for increased sedimentation of reservoirs and impacts on water quality (Smith *et al.*, 2011). In order to reduce potential post-fire sedimentation, reservoir managers have used, amongst various mitigation measures, erosion barrier sediment traps (Hobson *et al.*, 2004; Robichaud, 2005; deWolfe *et al.*, 2008) with varying levels of success. Considerable research has been conducted into the design and construction of sediment traps (Robichaud & Brown, 2002) and, to a lesser degree, the success of sediment traps (Robichaud *et al.*, 2000; Morris *et al.*, 2008; Robichaud, 2009; Fox, 2011). Reservoir managers are often limited in material resources and staff time, resulting in the need for sediment trap placement to be effective and efficient. Effective trap placement in a catchment requires an understanding of hillslope morphology and associated erosion processes in order to anticipate the key potential sediment sources and risks for mobilisation and delivery towards receiving water bodies.

Substantial research has been conducted on post-fire erosion processes (Shakesby & Doerr, 2006; Shakesby *et al.*, 2007; Shakesby, 2010). Predicting post-fire erosion hazards often involves applying erosion models (Fernandez *et al.*, 2005; Fox *et al.*, 2006; Miller *et al.*, 2011). Many of these models are not applicable to steep gradients above the angle of repose (Lamb *et al.*, 2011). Although the influence of slope gradient is well established in the theory of erosion (Sheridan *et al.*, 2003), the actual application of theory to the installation of post-fire sediment traps in relation to slope gradients needs further investigation. When applying post-fire erosion models, hillslope morphology is often simplified to incorporate the slope length and slope gradient (Merritt *et al.*, 2003; Fernandez *et al.*, 2005; Fox *et al.*, 2006). Rieke-Zapp & Nearing (2005) found that slope shape had a significant impact on rill patterns, sediment yield, and runoff production when modelling under controlled laboratory conditions. The irregular surface of the slope, such as the profile or the cross-slope (plan) curvature, is often overlooked (Rieke-Zapp & Nearing, 2005) or incorporated into modelling by dividing the topography into smaller morphological units (Di Piazza *et al.*, 2007).

There is a need to determine how hillslope morphology can influence post-fire surface erosion. Sediment trapping after the Mount Bold 2007 wildfire, in South Australia, provided a case study where hillslope erosion could be measured in relation to sediment trap success. The aim of this study was therefore to assess hillslope surface erosion in the context of post-wildfire sediment trapping in a catchment. The key components of the research involved: (a) quantifying hillslope surface change using erosion pins, terrestrial laser scanning and sediment traps after a 1 in 5 year rainfall event; (b) assessing the influence of slope gradient, slope length, cross-slope curvature, hillslope position and fire severity in relation to surface change; and (c) evaluating the success of sediment trap placement in relation to hillslope morphology.

STUDY AREA

The Mount Bold Reservoir reserve is located in the Southern Mount Lofty Ranges, on the Onkaparinga River, approximately 35 km southeast of Adelaide. The area lies in a temperate climatic zone with warm, dry summers and cool, wet winters. The mean annual rainfall at Mount Bold from 1939 to 2010 was 768 mm (Australian Bureau of Meteorology Mount Bold weather station, ID 023734; 35.07°S, 138.41°E; elevation 251 m). The geology of the area is comprised of the Bungarider subgroup containing the Stoneyfells quartzite and Woolshed Flat shale members, the Mundalio subgroup (Skillogalee dolomite), and the Emeroo subgroup, which contains quartzite, sandstone, dolomite and conglomerate (GSAA, 1962). The majority of the catchment has either a high or very high water erosion potential with soils being shallow to moderately deep acidic soils on rock (Soil & Land Program, 2007). Vegetation is typically *Eucalyptus* forest and woodlands, pine plantations or grasslands. The reservoir reserve contains large areas of remnant vegetation formations of Messmate Stringybark (*Eucalyptus obliqua*) associations including open forest, low open forest, woodland and low woodland (Pound, 2005).

A wildfire ignited by a suspected arsonist commenced on land adjoining the Mount Bold Reservoir on 10 January 2007 and burnt over 1500 hectares. Fire severities ranged from extreme (complete canopy and shrub defoliation with no leaves remaining) to low (leaf litter was partially consumed but the understorey was unburnt). The area had no recent fire activity with the last recorded fire being in the 1970s (EarthTech, 2004). Following the wildfire, the South Australian Water Corporation initiated emergency erosion mitigation works involving the installation of 53 sediment traps. Rain was observed nine days after the fire with a total of 46 mm falling over three days. Total annual rainfall at Mount Bold in 2007 was 760 mm (Australian Bureau of Meteorology Mount Bold weather station). Rainfall during the study period, from 10 January to 17 May 2007, totalled 225 mm, falling over 32 separate rain days. The most substantial rainfall event during this time occurred during the 82 hours to 30 April 2007. On the basis of rainfall intensity–frequency–duration analysis, data from the Houlgraves weather station (Australian Bureau of Meteorology Houlgraves weather station ID 023913; 35.05°S 138.44°E; elevation 250 m) located on the boundary of Mount Bold Reservoir, suggested that the 30 April rainfall event had a 1 in 5 year average recurrence interval.

Five hillslopes, sites A to E (Fig. 1), were selected for study, on the basis of all being located in the water reservoir catchment, within a distance of 5 km, and having similar elevation, geology and soil. Site A was distinctively different to the other sites due to the steep slopes ranging from 27° to 50°. Site B differed due to the convex/linear slope profile and linear cross-slope curvature, and the presence of occasional pine trees. Sites C and D had similar vegetation structural types and slope properties to each other, but site D was unburnt. All sites had been burnt during the 2007 wildfire, except for site D. Unlike the moderate to high severity fires affecting sites A to C, Site E was located within a sub-catchment that had been subjected to a very high severity fire. A sediment trap designed to mitigate sediment delivery to the water reservoir was located below each of the experimental hillslopes.

METHODS

Hillslope surface movement was assessed using erosion pins, terrestrial laser scanning and sediment traps. Erosion pins were used to monitor surface level changes following the wildfire at

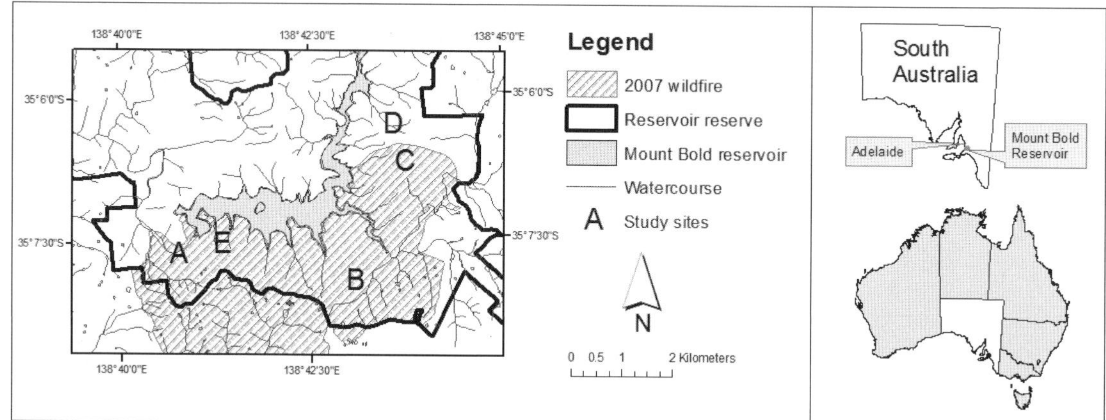

Fig. 1 Location map of the five study sites (A–E) at the Mount Bold reservoir in South Australia.

the five experimental hillslope locations. Targeting of erosion pin transects was based on the presence of an installed sediment trap, accessibility, permission from the water authority and sites of differing attributes including slope gradient, slope curvature, fire severity and vegetation type. In order to monitor differing hillslope positions the entire hillslope length was assessed, where possible, at 10-m intervals with replication by having two transects at each slope. Along four of the experimental hillslope profiles, sites B–E, two transect lines of pins were installed 10 m apart. To reduce operator bias when installing the pin, a tape measure was used to locate the pin entry point. At site A, one transect line was installed along the foot of the slope due to the steep slope gradient making the terrain inaccessible.

Installation of the erosion pins occurred between January and April 2007, and all pins were measured in May 2007. Pins were measured from the top to the ground surface using either callipers or metal rulers. A total of 126 marine grade stainless steel pins were installed. The pins were 4.7 mm in diameter and 500 mm long, except for 5 pins at site A that were 8 mm by 1500 mm. Erosion pins have previously been used to monitor hillslope erosion in temperate forests (Mackay *et al.*, 1984) and alpine areas in New South Wales, Australia (Smith & Dragovich, 2008), monsoonal savannah woodlands in Northern Territory, Australia (Russell-Smith *et al.*, 2006), moorlands in Yorkshire, UK (Imeson, 1971) and pine forest in Mexico (White & Wells, 1979). Haigh (1977) described numerous sources of data contamination when using erosion pins that included factors such as disturbance during establishment, influences on the pattern of soil erosion caused by the pin presence, trampling, vandalism, environmental variation, operator error and operator disturbance. The four erosion pins that had been disturbed during the study were not included in the analysed data set.

Terrestrial laser scanning (TLS) was used to model the surface level change at site A because slope steepness made the terrain inaccessible for erosion pins. Scans were conducted using a Maptek I-Site 4400LR in February and again in May 2007. The scanner is a time-of-flight pulsed rangefinder. Surface elevation models were created using Maptek I-Site studio software. Surface elevation change was modelled between February and May 2007.

Fifty-three sediment traps were constructed by the South Australian Water Corporation for emergency erosion mitigation. The vast majority of these traps were made from hay bales, star pickets and jute matting (Morris *et al.*, 2008). This study focuses on eight of the traps (Table 1) which were installed below the five experimental hillslope sites A to E. At site A, the sediment trap consisted of a 373-m long line of hay bales adjoining a recently installed road at the hillslope bottom (Fig. 2(a)). All other traps (Fig. 2(b)) were installed in dry channel positions using either hay bales or coir logs reinforced with star pickets and tensioned wire. Sediment volumes were determined using shovels, metal rulers and tape measures.

Fig. 2 Hay bale sediment traps at: (a) site A, trap 17, and (b) site C, trap 14a (images courtesy of Shayne Callis).

Table 1 Sediment trap description.

Site	Trap	Length (m)	Height (m)	Position	Material
A	17	373	0.5*	Foothill	490 hay bales, star pickets
B	21a	12	1.5	Channel	Jute matting, 14 hay bales, star pickets
C	14a	6	1	Channel	Jute matting, 12 hay bales, star pickets
C	15a	5	1	Channel	Jute matting, 12 hay bales, star pickets
D	Control	2.5	0.5	Channel	Jute matting, 2.5 coir logs, star pickets
E	1a	4	1	Channel	Jute matting, 8 hay bales, star pickets
E	1b	4	1	Channel	Jute matting, 8 hay bales, star pickets
E	1c	2	0.5	Channel	Jute matting, 2 hay bales, star pickets

*In a minority of sections the height was doubled to 1 m.

Statistical analyses of the results involved using both the net surface-level change and the absolute surface change. Absolute surface change gives a better indication of which sites were experiencing the most active sediment movement regime (Smith & Dragovich, 2008). A Kolmogorov-Smirnov statistic with a Lilliefors significance level was used to test for normality (n = 122, $p < 0.05$). The test confirmed that data for both the net surface-level change (statistic of 0.397) and the absolute surface change (statistic of 0.394) were not normally distributed. Non-parametric tests, including the Mann Whitney U and Kruskal-Wallis tests, were applied to the surface change data due to the non-normal data distribution. Correlations were assessed using the non-parametric Spearman rank correlation.

HILLSLOPE EROSION AND SEDIMENT TRAPPING

Hillslope surface movement

A comparison of mean net surface change between the burnt (A,B,C,E) and unburnt (D) hillslope sites yielded a net loss of –20.2 mm (SE ±12.1 mm) at the burnt sites and a net loss of –0.7 mm (SE ±0.4 mm) at the unburnt sites, a difference which is significant (Mann Whitney U test, Z = –3.171, $p < 0.05$). The sediment movement regime was more active across the burnt sites over the study period, with the absolute mean total surface change of 34.9 mm at the burnt sites exceeding the 1.3 mm at the unburnt site. The amount of sediment movement was higher at the burnt sites, with a third of the pin measurements exceeding ±10 mm of change compared to the unburnt site where no pins exceeded this surface change.

The mean net surface change was significantly different between the five study sites (Kruskal-Wallis test, X^2 (4, n = 122) = 13.175, $p < 0.05$). Burnt site A was noticeably different with change at 67% of the pins exceeding ±10 mm, and at 37% exceeding ±50 mm of change. Although to a lesser extent than site A, burnt site E also experienced substantial surface change with measurements exceeding both ±50 and ±10 mm change. Erosion pins were lost at both sites A and E. A 1500 mm erosion pin at Site A was entirely removed and lost due to a steep colluvial debris flow (pyrocolluviation). At site E, one of the 500 mm erosion pins was destabilized then washed 3 m downstream. Another pin at site E was bent by the force of material being transported during the April 2007 rain event. There was a trend for high burn severity sites to yield more sediment than moderate or very high severity sites.

Sediment yield reached a threshold when slope gradients were more than 1:2, equivalent to a slope angle of >26.6 degrees (Fig. 3(a)). There was a significant correlation between slope angle and absolute sediment movement (Spearman rank correlation, $R = 0.462$, $n = 122$, $p < 0.01$). When slopes exceeded 18 degrees, there was a nine-fold increase in mean absolute sediment movement. There was also significant correlation between slope length and absolute sediment movement (Spearman rank correlation, $R = 0.285$, $n = 122$, $p < 0.01$). Comparison of mean absolute sediment movement (Fig. 3(b)) between concave, (136.9 mm, SE ±64.0), convex (41.9 mm, SE ±12.4) and linear (5.4 mm, SE ±0.7) cross-slope curvatures were significantly different (Kruskal-Wallis test, X^2 (2, $n = 122$) = 20.554, $p < 0.05$). The greatest mean surface change occurred at foothill slope positions (–45.1 mm, SE ±25.3) or within drainage lines (–15.8, SE ±10.2). When the steepest location, site A, was not included in the statistical analysis, the foothill position mean surface change altered to 2.3 mm (SE ±2.3).

A direct comparison of burnt Site A with the other experimental sites is limited by differing erosion pin configuration. To account for this difference, surface elevation models were created using the terrestrial laser scanning. The surface elevation modelling highlighted that the greatest surface change (>1 m) generally occurred in the concave cross-slope area of the hillslope (Fig. 4). Immediately following the fire, initial sediment movement occurred as dry ravel then subsequent rainfall events resulted in colluvial debris flows. The 1 in 5 year rainfall event caused an entire 1.5 m erosion pin to be dislodged and subsequently lost within the debris flow. The terrestrial laser scanning also modelled the surface change at this erosion pin location to be >1 m.

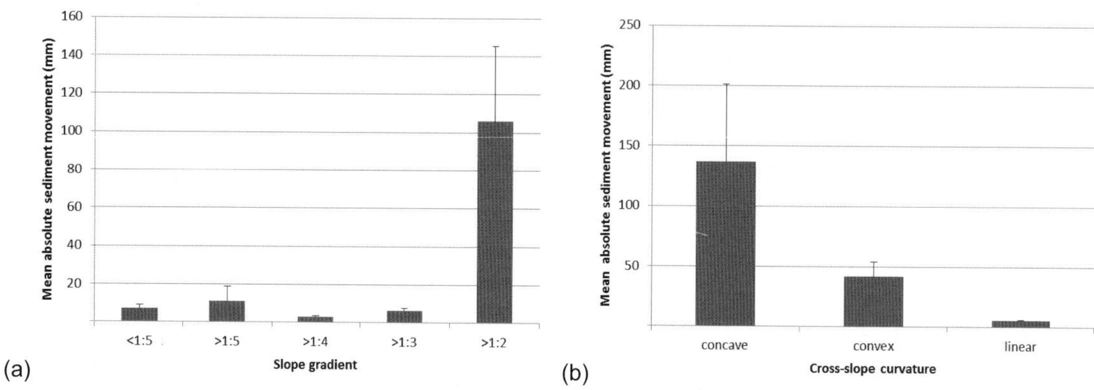

Fig. 3 Mean absolute sediment movement (+SE error bars) for: (a) slope gradient, and (b) mean absolute sediment movement for cross-slope curvature.

Sediment traps

Seventeen of the 53 sediment traps installed at Mount Bold partially failed to capture the moving sediment and debris following the April 2007 rainfall event (Morris *et al.*, 2008). Three of the five study sites (Table 2) captured sediment; however, the trap size was still not sufficient to capture all of the sediment moved. At site A, the 370-m long hay bale trap was often destroyed or breached below concave cross-slope curvatures where water flow converged. At site E, the hay bale traps were completely destroyed by the velocity of the water and the force of the transported sediment

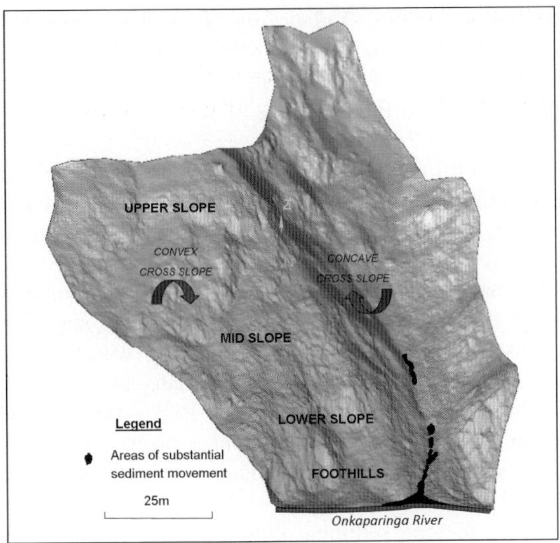

Fig. 4 Hillslope digital elevation surface model of sediment movement at site A between February and May 2007, using terrestrial laser scanning.

Table 2 Summary of trap success and the hillslope surface level change for each study site.

Study site	Trap failure	SV (m^3)	MAS (mm) (SE)	S (mm) (SE)	Fire severity	Slope degree	Slope gradient
A	Yes	>103	105.9 (39.2)	–76.7 (41.5)	M–H	27–50	1:2–1:1
E	Yes	>2.3	9.6 (3.4)	–2.5 (3.9)	VH	3–24	1:19–1:2
C	Yes	>8	7.9 (1.8)	2.6 (2.6)	M–H	0–22	1:57–1:2
B	No	0	4.9 (0.9)	4.4 (1.0)	H	0–25	1:57–1:2
D	No	0	1.3 (0.2)	–0.7 (0.4)	U	2–18	1:29–1:3

SV: Trapped sediment volume (minimum due to trap failures); MAS: Mean absolute surface level change; S: Mean surface level change; VH: very high; H: high; M: moderate; U: unburnt; (SE) Standard error.

and debris. The remaining two study sites (Table 2) contained traps that did not capture any quantifiable amounts of sediment.

The absolute mean surface change measured at site A was 10 times greater than at all other sites (Table 2). The sediment volume captured at site A was 12 times larger than the volume of sediment captured at the other sites. Sites A, D and E all had negative surface changes implying that surface erosion was the dominant sediment mobilisation process. It was only at Site A that negative values exceeded –75 mm. The sediment traps overflowed at sites A, C and E, whereas site B and the control site D did not sequester any quantifiable volumes of sediment.

DISCUSSION

Knowledge and appreciation of hillslope morphology and associated erosion processes can assist land managers in the targeting of mitigation measures for trapping mobilised sediment post-fire. In the case of Mount Bold, the capture of sediment following a 1 in 5 year rainfall event varied depending on slope properties. Similar to most post-fire studies (e.g. Shakesby & Doerr, 2006; Sheridan *et al.*, 2007; Moody & Martin, 2009) surface movement increased substantially in areas that had been subjected to fire. The expected influence of fire severity on sediment movement was not detected from the data assembled, possibly due to the influence of slope properties at sites A and B.

Slope gradient and the cross-slope curvature may have influenced the differences in surface change between sites A and B. Site A was extremely steep, with the slope ranging from 27 to 50 degrees. The approximate angle of repose, where material may slide down the surface, is 34 degrees for dry sandy soil. Post-fire dry ravel gravitational movement, as observed by Lamb *et al.* (2011), was also noted at site A. In contrast, other sites such as site B had gentler slopes (0–25 degrees) and a linear cross-slope curvature that reduced the converging nature of surface water flow. Site A had concave cross-slope curvature that resulted in converging water flows. Our erosion pin results and TLS models indicated that surface change was greater at concave rather than linear cross-slope curvatures. This result contrasts with those of Rieke-Zapp & Nearing (2005) who reported that laboratory linear slopes generated the highest mean sediment yield. Along site A, the sediment traps below linear slopes remained intact, whereas below concave cross-slope curvatures the traps were generally breached.

Hay bale traps were sufficient to capture moving sediment during regular rainfall events at Mount Bold. When the average rainfall intensity–frequency–duration increased to a 1 in 5 year event, many of the traps were breached. Hobson *et al.* (2004) at Little Para Reservoir in Adelaide, South Australia, and Robichaud *et al.* (2008) in western Montana, USA, also found that natural rainfall events caused the sediment-laden runoff to overtop hay bales. At Mount Bold site A, sediment overtopped the trap at numerous locations resulting in substantial material reaching the Onkaparinga Creek within the water reservoir. At the foothill of steep slopes with gradients of ≥1:2 or within concave drainage lines, hay bale traps are unlikely to survive. Land managers need either to implement alternative trap designs such as rock gabions or to accept the likely outcome of increased reservoir sedimentation below steep slopes that require alternative management options. Alternative mitigation strategies such as mulches, seeding, geotextile bags and silt fencing can be combined with hay bales to improve the trapping efficiency (Robichauld, 2009). Mulching has been shown to be effective at reducing post-fire erosion; however, it is relatively expensive (Bautista *et al.*, 2009). As the Mount Bold study is limited to areas with specific soil types and *Eucalyptus* vegetation, further research is needed into sediment trapping on hillslopes with slope gradients ≥1:2 and with differing soils and vegetation. Incorporating slope profile and cross-slope curvature into post-fire mitigation assessment and erosion modelling also requires further investigation.

CONCLUSION

Substantial sediment movement resulting in surface changes of greater than 1 m occurred after a 1 in 5 year rainfall event at Mount Bold Reservoir reserve, and this resulted in the failure of numerous sediment traps implemented as part of the mitigation strategy. In this rainfall event, fire severity (between moderate to very high), as a factor influencing trap effectiveness, was overshadowed by the importance of slope properties, even though the presence of fire was necessary to trigger the considerable sediment mobilization and delivery observed. Higher slope gradients and longer slopes contributed to greater sediment transfer, with the largest surface change occurring in footslope positions. On the basis of the data assembled by this study, a threshold for substantial increase in sediment yield was identified at slope gradients of 1:2. Concave cross-slope curvature was also associated with significantly larger amounts of sediment movement. Incorporating the influence of hillslope morphology may be useful when using sediment traps in tandem with other measures, such as mulching and seeding. Concentrated mitigation efforts could focus on the concave cross-slope curvature of the slope. The successful placement of sediment traps and other mitigation strategies to protect water reservoirs requires an understanding of hillslope morphology and the ways in which it influences and controls erosion and delivery processes.

Acknowledgements Thanks are extended to staff from the South Australia Water, especially Shayne Callis, Monique Blason and Bert Eerden. Terrestrial laser scanning was conducted by James Moncrieff from Maptek Pty. Linton Johnson from the Bureau of Meteorology analysed the rainfall–intensity–frequency duration analysis data. Guidance was provided from PhD supervisors Ross Bradstock and Meredith Henderson. Funding from the Bushfire Cooperative Research Centre is acknowledged. Thanks are also extended to the anonymous reviewers for their helpful suggestions.

REFERENCES

deWolfe, V. G., Santi, P. M., Ey, J. & Gartner, J. E. (2008) Effective mitigation of debris flows at Lemon Dam, La Plata County, Colorado. *Geomorphol.* 96(3-4), 366–377, doi:10.1016/j.geomorph.2007.04.008.

Di Piazza, G. V., Di Stefano, C. & Ferro, V. (2007) Modelling the effects of a bushfire on erosion in a Mediterranean basin. *Hydrol. Sci. J.* 52(6), 1253–1270, doi: 10.1623/hysj.52.6.1253.

Earth Tech (2004) Mount Bold Reservoir Reserve, including Clarendon Weir Land Management Plan. Job 6503011-R004, South Australian Water Corporation. 228p.

Fernandez, S., Marquinez, J. & Menendez Duarteb, R. (2005) A susceptibility model for post wildfire soil erosion in a temperate oceanic mountain area of Spain. *Catena* 61, 256–272, doi: 10.1016/j.catena.2005.03.006.

Fox, D., Berolo, W., Carrega, P. & Darboux, F. (2006) Mapping erosion risk and selecting sites for simple erosion control measures after a forest fire in Mediterranean France. *Earth Surf. Processes Landf.* 31(5), 606–621, doi:10.1002/esp.1346.

Fox, D. M. (2011) Evaluation of the efficiency of some sediment trapping methods after a Mediterranean forest fire. *J. Environ. Manage.* 92(2), 258–265, doi:10.1016/j.jenvman.2009.10.006.

GSAA (1962) Barker 1:250,000 Mapsheet. S.A. Geological Atlas Sheet 1 54-13 Zones 5 & 6. Geological Survey of South Australia.

Haigh, M. J. (1977) The use of erosion pins in the study of slope evolution. *Brit. Geomorph Res Grp Tech Bull.* 18, 31–48.

Hobson, P., Hackney, P. & Brookes, J. (2004) Effects of bushfire on water quality in Little Para Reservoir, South Australia. *AWA Branch Operators Conference,* Adelaide.

Imeson, A. C. (1971) Heather burning and soil erosion on North Yorkshire Moors. *J. Appl. Ecol.* 8, 537–542.

Lamb, M. P., Scheingross, J. S., Amidon, W. H., Swanson, E. & Limaye, A. (2011) A model for fire-induced sediment yield by dry ravel in steep landscapes. *J. Geophy. Res.-Earth Surface* 116, F03006, 13 PP, doi: 10.1029/2010JF001878.

Mackay, S. M., Long, A. C. & Chalmers, R. W. (1984) Erosion pin estimates of soil movement after intensive logging and wildfire. In: *Drainage Basin Erosion and Sedimentation: Conference and Review Papers 2* (ed. by R. J. Loughran), 15–22 (University of Newcastle, Australia).

Merritt, W. S., Letcher, R. A. & Jakeman, A. J. (2003) A review of erosion and sediment transport models. *Environ. Modeling & Software* 18(8-9), 761–799, doi:10.1016/S1364-8152(03)00078-1.

Miller, M. E., MacDonald, L. H., Robichaud, P. R. & Elliot, W. J. (2011) Predicting post-fire hillslope erosion in forest lands of the western United States. *Int. J .Wildland Fire* 20(8), 982–999, doi:10.1071/WF09142.

Moody, J. A. & Martin, D. A. (2009) Synthesis of sediment yields after wildland fire in different rainfall regimes in the western United States. *Int. J .Wildland Fire* 18(1), 96–115, doi:10.1071/WF07162.

Morris, R. Calliss S., Frizenschaf, J., Blason, M., Dragovich, D., Henderson, M. & Ostendorf, B. (2008) Controlling sediment movement following bushfire – a case study in managing water quality, Mount Bold, South Australia. Water Downunder, Adelaide, Australia, 1937–1947. Engineers Australia, Modbury.

Pound, L. (2005) A biological survey of flora and fauna at Mount Bold reservoir reserve. Nature Conservation Society of South Australia, Adelaide, Department of Water, Land and Biodiversity Conservation, South Australia.

Rieke-Zapp, D. H. & Nearing, M. A. (2005) Slope shape effects on erosion: A laboratory study. *Soil Sci. Soc. Am. J.* 69(5), 1463–1471, doi: 10.2136/sssaj2005.0015.

Robichaud, P. R. (2005) Measurement of post-fire hillslope erosion to evaluate and model rehabilitation treatment effectiveness and recovery. *Int. J. Wildland Fire* 14(4), 475–485, doi:10.1071/WF05031.

Robichaud, P. R. (2009) Using erosion barriers for post-fire stabilization. In: *Fire Effects on Soils and Restoration Strategies* (ed. by A. Cerda & P. R. Robichaud), 337–352. Science Publishers, Oxford.

Robichaud, P. R., Beyers, J. L. & Neary, D. G. (2000) Evaluating the effectiveness of postfire rehabilitation treatments. RMRS-GTR-63.

Robichaud, P. R. & Brown, R. E. (2002) Silt fences: An economical technique for measuring hillslope soil erosion. USDA Forest Service Rocky Mountain Research Station.

Robichaud, P. R., Pierson, F. B., Brown, R. K. & Wagenbrenner, J. W. (2008) Measuring effectiveness of three postfire hillslope erosion barrier treatments, western Montana, USA. *Hydrol. Processes* 22(2), 159–170. doi:10.1002/hyp.6558.

Russell-Smith J., Yates, C. & Lynch, B. (2006) Fire regimes and soil erosion in north Australian hilly savannas. *Int. J .Wildland Fire* 15(4), 551–556. doi:10.1071/WF05112.

Shakesby, R. A. (2010) Post-wildfire soil erosion in the Mediterranean: review and future research directions. *Earth-Science Rev.* 105(3-4), 71–100, doi: 10.1016/j.earscirev.2011.01.001.

Shakesby, R. A. & Doerr, S. H. (2006) Wildfire as a hydrological and geomorphological agent. *Earth-Science Rev.* 74(3-4), 269–307, doi:10.1016/j.earscirev.2005.10.006.

Shakesby, R. A., Wallbrink, P. J., Doerr, S. H., English, P. M., Chafer, C. J., Humphreys, G. S., Blake, W. H. & Tomkins, K. M. (2007) Distinctiveness of wildfire effects on soil erosion in south-east Australian eucalypt forests assessed in a global context. *Forest Ecol. and Manage.* 238(1-3), 347–364, doi:10.1016/j.foreco.2006.10.029.

Sheridan, G. J., Lane, P. N. J. & Noske, P. J. (2007) Quantification of hillslope runoff and erosion processes before and after wildfire in a wet Eucalyptus forest. *J. Hydrol.* 343(1-2), 12–28. doi:10.1016/j.jhydrol.2007.06.005.

Sheridan, G. J., So, H. B. & Loch, R. J. (2003) Improved slope adjustment functions for soil erosion prediction. *Aust. J. Soil Res.* 41(8), 1489–1508. doi:10.1071/SR02029.

Smith, H. G. & Dragovich, D. (2008) Post-fire hillslope erosion response in a sub-alpine environment, south eastern Australia. *Catena* 73, 274–285. doi:10.1016/j.catena.2007.11.003.

Smith, H. G., Sheridan, G. J., Lane, P. N. J., Nyman, P. & Haydon, S. (2011) Wildfire effects on water quality in forest catchments: a review with implications for water supply. *J. Hydrol.* 396(1-2), 170–192. doi:10.1016/j.jhydrol.2010.10.043.

Soil & Land Program (2007) Land and Soil Spatial Data for Southern South Australia – GIS format [CD Rom].

White, W. D. & Wells, S. G. (1979) Forest-fire devegetation and drainage basin adjustments in mountainous terrain. In: *Adjustments of the Fluvial System* (ed. by D. D. Rhodes & G. P. Williams), 199–223. Kendall/Hunt Publishing, Iowa, USA.

Effects of flow regime on stream turbidity and suspended solids after wildfire, Colorado Front Range

SHEILA F. MURPHY[1], R. BLAINE McCLESKEY[1] & JEFFREY H. WRITER[1,2]

1 *US Geological Survey, 3215 Marine Street, Boulder, Colorado 80303, USA*
sfmurphy@usgs.gov

2 *University of Colorado, Department of Civil, Environmental, and Architectural Engineering, 428 UCB, Boulder, Colorado 80309, USA*

Abstract Wildfires occur frequently in the Colorado Front Range and can alter the hydrological response of watersheds, yet little information exists on the impact of flow regime and storm events on post-wildfire water quality. The flow regime in the region is characterized by base-flow conditions during much of the year and increased runoff during spring snowmelt and summer convective storms. The impact of snowmelt and storm events on stream discharge and water quality was evaluated for about a year after a wildfire near Boulder, Colorado, USA. During spring snowmelt and low-intensity storms, differences in discharge and turbidity at sites upstream and downstream from the burned areas were minimal. However, high-intensity convective storms resulted in dramatic increases in discharge and turbidity at sites downstream from the burned area. This study highlights the importance of using high-frequency sampling to assess accurately wildfire impacts on water quality downstream.

Key words wildfire; water quality; turbidity; Colorado Front Range; Fourmile Canyon fire; flow regime; convective storms

INTRODUCTION

Wildfire size, fire severity and length of fire season have increased in recent years, largely due to effects from climate variability, including extreme droughts, forest disease outbreaks, and changes in precipitation patterns (Westerling *et al.*, 2006; Holden *et al.*, 2007). Wildfires can drastically alter the hydrological and soil properties of watersheds (Moody & Martin, 2001), leading to increased flooding, erosion, and sediment loads. Large erosion events, such as those caused by wildfire, are short-lived on geological time scales, but can dominate the long-term sediment yield response (Kirchner *et al.*, 2001). Wildfires can impair water quality and are a major disturbance in many terrestrial and aquatic ecosystems. Studies of surface waters after wildfire have reported increases in turbidity, nutrients, organic carbon, sulfate, major ions and trace metals (e.g. Tiedemann *et al.*, 1979; Gresswell, 1999). Water quality effects from wildfire vary substantially depending on several factors including, amongst others, the proportion of watershed burned, steepness of watershed slopes, geology, and post-fire precipitation type, timing and intensity (Beschta, 1990; Neary *et al.*, 2005).

The Fourmile Canyon Fire burned 2600 hectares and destroyed more than 160 homes in Boulder County, Colorado, USA in September 2010. The wildfire burned 23% of the Fourmile Creek Watershed (Fig. 1). As a result of high winds and rapid movement of the fire front, burn severity ranged from low to severe (Keeley, 2009; Fourmile Emergency Stabilization Team, 2010). The Fourmile Creek watershed consists largely of steep, rugged terrain and the wildfire has left the area at risk of substantial erosion, including debris flows (Ruddy *et al.*, 2010). The watershed also contains abandoned waste rock and tailings piles from historical mining, many of which have been exposed by the wildfire. Fourmile Creek discharges to Boulder Creek about 3 km upstream from the city of Boulder, Colorado. Fourmile Creek and Boulder Creek are sources for water supply for local communities. Several other wildfires have occurred in this region in the past 25 years (Fig. 1). Therefore, an in-depth study of the hydrological and hydrochemical responses of this watershed was warranted to assess how watershed processes are affected and to determine the increase in risks of flooding, debris flows and water quality impairment.

The flow regime in this watershed, as is typical in the Colorado Front Range and many watersheds in the western USA, is characterized by base-flow conditions during much of the year, with increased runoff during spring snowmelt and high-intensity summer convective storms.

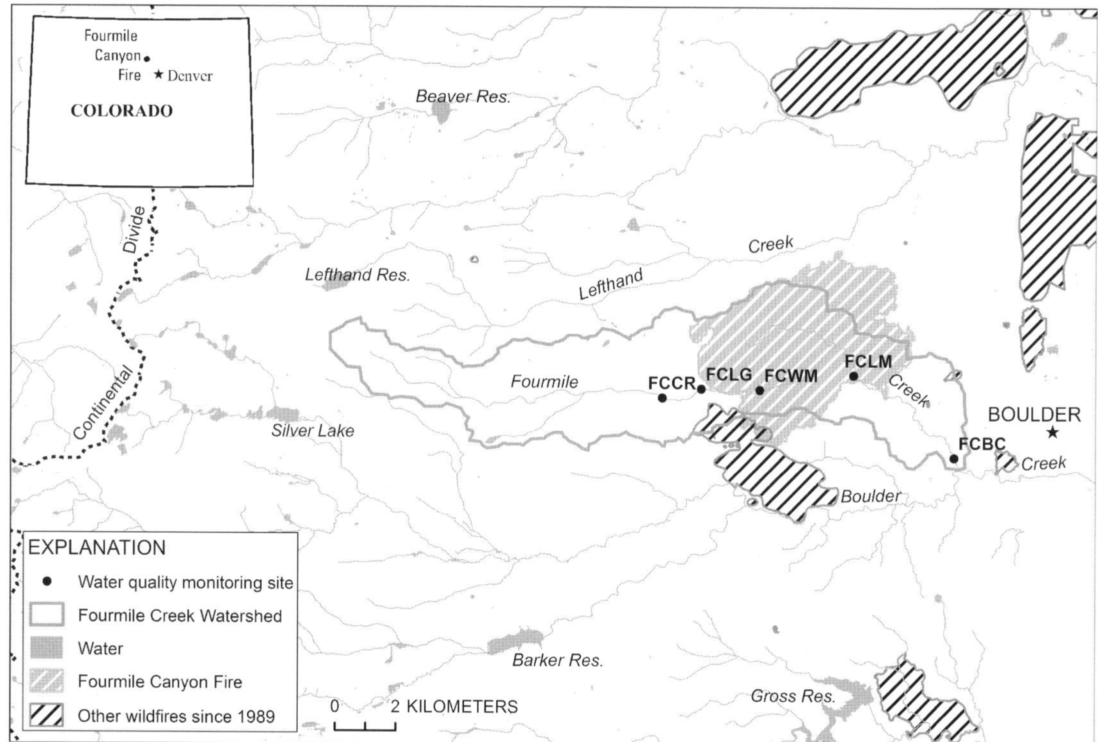

Fig. 1 Map showing the study location (wildfire areas from Boulder County Land Use Department).

Water quality is closely tied to this annual regime, yet little information exists on the impact of snowmelt and convective storms on water quality after wildfire. Weekly or monthly water-quality sampling is not sufficient to capture effectively the relation between watershed hydrology and streamwater chemistry (Kirchner *et al.*, 2004), but the remote locations of many wildfires make frequent manual sampling difficult. This study focused on high-frequency water-quality sampling at several locations in the Fourmile Creek watershed for the purpose of understanding the coupling of hydrological and chemical processes, and to evaluate the effects of snowmelt runoff and summer convective storms on water quality. Specifically, this paper focuses on post-wildfire turbidity and total suspended solids. The export of carbon and nitrogen is discussed in Writer *et al.* (2012).

METHODS

Precipitation data from raingauges within and surrounding the burned area were obtained from the Urban Drainage and Flood Control District (UDFCD) and the National Atmospheric Deposition Program (NADP) (Fig. 1). The gauges operated by UDFCD (2011) are 1-mm tipping-buckets and the tips were summed to obtain daily precipitation, storm totals and maximum 30-minute rain intensity (which was converted to units of mm h^{-1} and is abbreviated here as I_{30}). Daily precipitation at five raingauges in, or near to, the burned area was averaged to estimate average daily precipitation in the region for the period June–September (these raingauges are not designed to record snow, which commonly occurs from October to May). Data from the NADP Sugarloaf station is recorded by an ETI Total Precipitation Gauge NOAH IV, which is an all-weather precipitation gauge that weighs the entire contents of an internal collection container. Both daily and 15-minute precipitation data are reported (NADP, 2011) from this gauge. These daily data were used to estimate precipitation in the region for the period October–May. Daily minimum temperature was also obtained from the NADP Sugarloaf station (NADP, 2011).

Five sites on Fourmile Creek were monitored for water quality and quantity for over a year following the wildfire (Fig. 1), and four of these were instrumented to obtain continuous discharge data. Historical (1947–1953, 1983–1995) mean daily discharge and peak discharge data from USGS stream-gauging station 06727500 (located 0.4 km upstream from Boulder Creek at site FCBC; Fig. 1) were obtained from the US Geological Survey (USGS, 2011). A stream-gauging station was re-installed at this location by the USGS Colorado Water Science Center in 2011. An additional station (06727410) was installed on Fourmile Creek at Logan Mill Road about 5.8 km upstream of Boulder Creek (at the FCLM site) in 2011. These two stations reported discharge every five minutes from 1 April to 30 September 2011. Vented water-level loggers (Global Water, model WL16U) were deployed at sites FCCR and FCWM (and FCLM and FCBC prior to the installation of USGS stream-gauging stations) to monitor stage. Prior to deployment, the water-level loggers were calibrated by measuring instrument response at 0.1-m intervals for water depths from 0 to 1.1 m. The water level loggers were housed within 1.5-m long sections of 5-cm diameter, perforated PVC pipe anchored to the streambed with rebar. Data were downloaded from the instruments at approximate 30-day intervals. Stream discharge was measured periodically with a pygmy meter using standard US Geological Survey protocols (Rantz *et al.*, 1982), and these measurements were used to generate a rating curve for each site, and to subsequently estimate stream discharge from stage readings.

The four sites monitored for continuous stream discharge (FCCR, FCWM, FCLM, and FCBC; Fig. 1) were also instrumented with conductivity loggers from Onset Corporation (Hobo data logger U24-001) to monitor continuously electrical conductivity and temperature. Water samples were collected monthly during base-flow conditions and twice weekly during snowmelt runoff. Electrical conductivity (Amber Science, Model 2052) and pH and temperature (Orion, 3-Star pH/temperature meter) were measured in the stream. Water samples were collected in pre-cleaned 1-L Teflon bottles and immediately returned to the USGS National Research Program laboratory in Boulder, Colorado for filtering, preservation, and measurement of turbidity (within 4 hours). Water samples were collected at more frequent intervals using automatic samplers (ISCO, models 6700 and 6712) during, or immediately after, substantial precipitation events (12 October 2010; 15–17 April 2011; 18–20 May 2011; 19–21 June 2011; 7–8 July 2011; 13–14 July 2011; and 7 September 2011) at sites FCCR, FCWM, FCLM, and FCBC. Automatic samplers were either triggered manually or when water level reached a pre-set stage. These samples were collected in pre-cleaned 1-L polyethylene sampling bottles and were delivered to the laboratory for measurement of electrical conductivity and turbidity, filtration, and preservation within 48 hours. The turbidity of unfiltered samples was measured using a Hach 2100Q portable turbidimeter, which was calibrated with 20, 100 and 800 nephelometric turbidity unit (NTU) formazin standards provided by the manufacturer (Hach Company, 2010). If the turbidity of the sample exceeded the range of the instrument (1000 NTU), the sample was diluted with de-ionized water, agitated, and re-analysed. A subset of samples was analysed for total suspended solids (TSS). The mass of dried sediment suspended on a 0.4-micrometer pore size filter, plus the mass of sample passed through the filter were recorded, and the mass of water sample was converted to volume assuming a water density of 1 g cm^{-3} (very turbid samples may have a greater density, but by comparing sample volume to mass, the maximum error related to this assumption was determined to be less than 7%). The relation between turbidity and TSS was calculated using a simple linear regression, as described below. Filtered samples were analysed for dissolved organic carbon, nutrients, major cations and anions, plus metals (McCleskey *et al.*, 2012).

RESULTS AND DISCUSSION

The historical (1947–1953, 1983–1995) mean daily discharge of Fourmile Creek at FCBC ranged from 0.025 to 0.96 m^3 s^{-1} (USGS, 2011). Discharge in April and the first half of May 2011 was lower than the historical mean, as a result of low snowfall (Fig. 2(a)). Discharge rose sharply in mid-May after minimum air temperatures rose above 0°C and two substantial precipitation events occurred (Fig. 2(b), Table 1). Several large storms in June and July led to discharge values that were higher than the historical daily mean. A convective storm on 13 July 2011 resulted in a peak discharge of 23 $m^3 s^{-1}$ at site FCLM, downstream from the burned area (Fig. 3(a)). This storm conveyed a substantial amount of sediment to Fourmile Creek, which moved through the stream system over the next few weeks, causing uncertainty in discharge values in late July to early August as sediment moving through the system modified the stream bed (and rating curve).

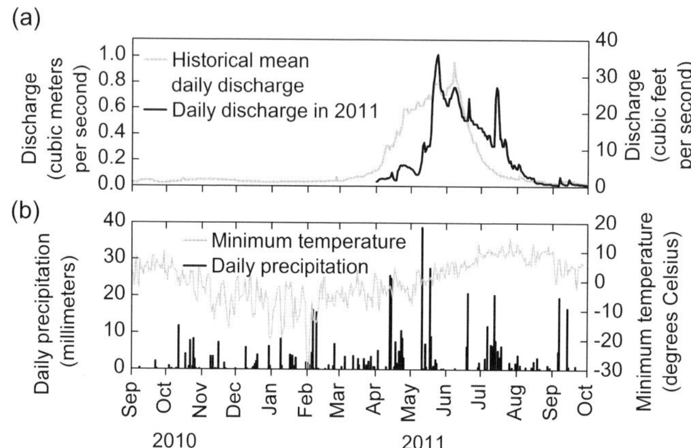

Fig. 2 (a) The historical mean daily discharge (1947–1953, 1983–1995) and mean daily discharge in 2011 of Fourmile Creek (at USGS streamflow-gauging station 06727500; USGS, 2011). (b) Precipitation and minimum daily air temperature, 1 September 2010 to 30 September 2011 (1 September 2010–31 May 2011 precipitation from Sugarloaf station, National Atmospheric Deposition Program (2011); 1 June–1 October 2011 precipitation from average of five Urban Drainage and Flood Control District raingauges in or near the burned area; temperature from Sugarloaf station (NADP, 2011)).

Table 1 Storms with daily precipitation totals greater than 10 mm, September 2010–September 2011.

Date	Type of precipitation	Total precipitation (mm)	Maximum I_{30}^* ($mm\ h^{-1}$)
12-Oct-10	rain	14	4
6-Feb-11	snow	16	--
9-Feb-11	snow	16	--
13-Apr-11	snow	26	--
14-Apr-11	snow	25	--
23-Apr-11	snow	11	--
24-Apr-11	snow	9	--
11-May-11	mix	39	--
18-May-11	mix	28	--
19-May-11	mix	9	--
19-Jun-11	rain	6	
20-Jun-11	rain	21	16 †
7-Jul-11	rain	12	46
13-Jul-11	rain	20	40
7-Sep-11	rain	20	8
14-Sep-11	rain	17	8

Dates with <10 mm included if part of the same storm as previous or following day; from Sugarloaf NADP station for dates with snow or mixed; average of 5 stations closest to the burned area for dates with rain only. *I_{30}, maximum 30-minute rainfall intensity at any of five gauges nearest the burned area (calculated for rain-only events)]

† Maximum value on 19 or 20 June 2011.

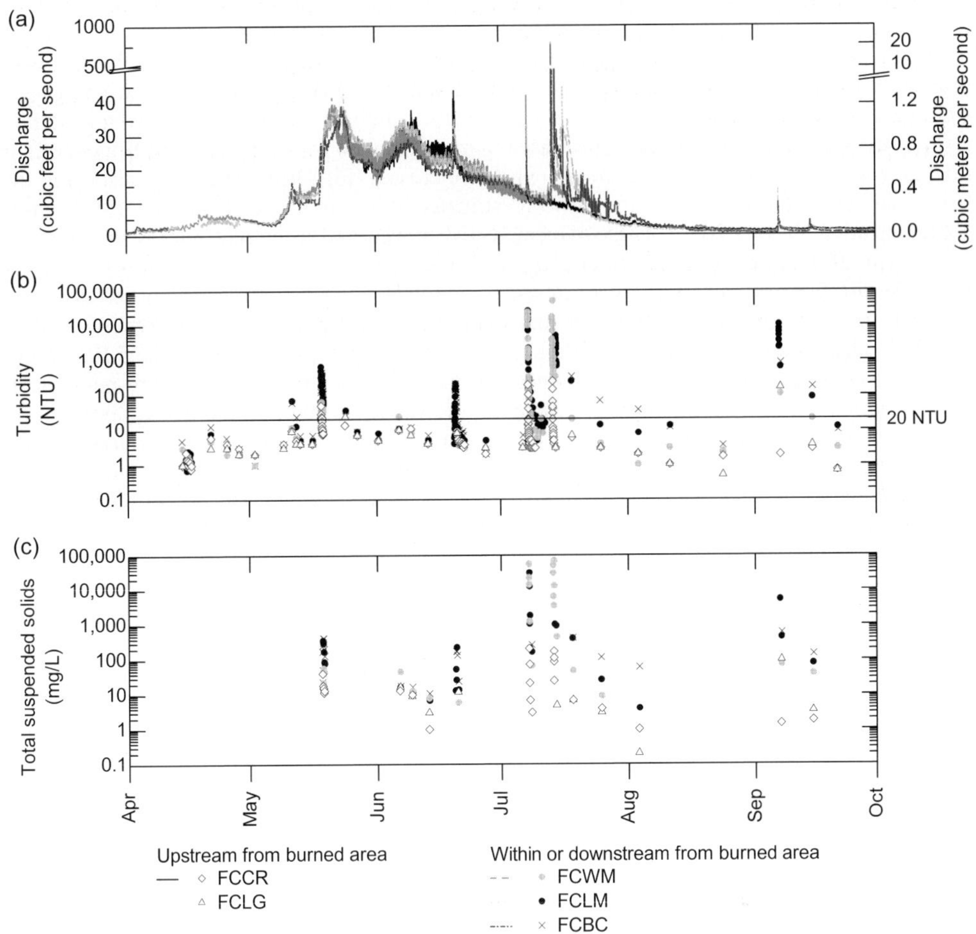

Fig. 3 Discharge, turbidity, and TSS measured in Fourmile Creek, 1 April to 30 September 2011.

Discharge at four locations along Fourmile Creek was similar during this study (Fig. 3(a)), suggesting that most of the water in the creek at all sampling locations is derived from upstream. Fourmile Creek has few major tributaries – the most substantial is the ephemeral Gold Run (Fig. 1) – and snowfall is greater in the upper reaches of the watershed. Summer rain storms varied in extent, total precipitation and intensity, and thus affected discharge at the monitoring sites differently. A large frontal storm on 19–20 June 2011 delivered an average of 27 mm precipitation to the five raingauges in, or near, the burned area (Table 1). This was the largest storm total in June/July 2011; however, this storm had a relatively low rainfall intensity (maximum I_{30} of 16 mm h^{-1}; Table 1). This storm also had a relatively consistent intensity across the Fourmile Creek watershed. In contrast, convective storms on 7 and 13 July 2011 had lower precipitation totals (average values of 12 and 20 mm, respectively), but much higher rainfall intensities (maximum I_{30} values of 46 and 40 mm h^{-1}, respectively). These two convective storms were much smaller in extent, and were centred on different areas of the Fourmile Creek watershed. The 7 July 2011 storm was centred upstream from and within the western part of the burned area, near sampling sites FCCR, FCLG, and FCWM. Prior to the storm, discharge at all Fourmile Creek sites was between 0.29 and 0.34 $m^3 s^{-1}$

(Fig. 3(a)). During the storm, discharge at FCCR (located upstream from the burned area) increased to 0.4 $m^3 s^{-1}$ (49% increase), while discharge at FCWM and FCLM (within or downstream from the burned area) increased to 1.2 $m^3 s^{-1}$ (313% increase) and 0.93 $m^3 s^{-1}$ (175% increase), respectively (Fig. 3(a)). The 13 July 2011 storm was centred on Gold Run, which discharges to Fourmile Creek between sites FCWM and FCLM (Fig. 1). Pre-storm discharge at all sites was between 0.26 and 0.28 $m^3 s^{-1}$. During the storm, discharge at FCCR increased to 0.35 $m^3 s^{-1}$ (36% increase), discharge at FCWM increased to more than 1.0 $m^3 s^{-1}$ (at least 281% increase) and discharge at FCLM increased to 23 $m^3 s^{-1}$ (8100% increase) (Fig. 3(a)). Maximum discharge from Gold Run during the 13 July storm has been estimated at 8.2 $m^3 s^{-1}$ (R. D. Jarrett, USGS, written communication, 20 July 2011). These two convective storms had recurrence intervals of about 2 to 5 years (for 30-minute intervals; Miller *et al.*, 1973), i.e. have a 20–50% chance of occurring each year. Despite the relatively low storm totals, the 13 July 2011 event resulted in an increase in discharge that was about three times greater than ever recorded on Fourmile Creek (7.3 $m^3 s^{-1}$; USGS, 2011). Wildfires reduce the threshold precipitation intensity at which overland flow occurs; a 5-year recurrence interval storm in a burned forest can produce the same runoff response as that of an unburned forest during a 30-year recurrence interval storm (Wondzell & King, 2003).

During periods of no precipitation, including during snowmelt runoff, turbidity and TSS at all sites were low, and similar from site to site (Fig. 3). During precipitation events, turbidity and TSS at all sites typically increased, but turbidity and TSS at sites within and downstream from the burned area showed greater increases than upstream sites; dissolved organic carbon showed similar trends (Writer *et al.*, 2012). These increases were substantially greater during the high-intensity convective storms in July; turbidity and TSS values were as high as 50 000 NTU and 68 000 milligrams per litre (mg L^{-1}), respectively, within the burned area. These TSS values fall within a range of maximum reported TSS concentrations measured globally during stormflows in the first year after fire (11 to 500 000 mg L^{-1}; Smith *et al.*, 2011). Due to the large amount of sediment delivered to Fourmile Creek, particularly downstream from Gold Run, turbidity values downstream from the burned area remained elevated for several weeks after the storms – even when no precipitation occurred – as base flows continued to transport sediment through the system. A rain event on 7 September 2011, which had an I_{30} of only 8 mm h^{-1} (Table 1), remobilized stream sediment and led to turbidity values of 10 000 NTU at site FCLM (automatic samplers at other sites did not trigger during this event). Our findings that post-wildfire erosion was most significant during high-intensity storm events (in this case, 10 months after the wildfire) are similar to those of Benavides-Solorio & MacDonald (2005), who found that 90% of the sediment delivered after wildfire was generated by summer convective storms. These high-intensity storms led to substantial erosion in the study watershed, despite post-fire stabilization efforts that included aerial straw mulching and seeding of native plants (Boulder County, 2012). It was beyond the scope of this study to determine if these mitigation efforts were effective in decreasing erosion. Straw mulching can be significantly effective at reducing sediment yield, but effectiveness can be decreased due to uneven application and redistribution related to wind conditions, hillslope steepness and standing trees (Robichaud *et al.*, 2010), all of which are factors in the Fourmile Creek watershed.

The values for TSS showed a strong correlation with turbidity (Fig. 4). The relation for all samples was TSS = 1.5 × turbidity + 80, with an r^2 of 0.86 (Fig. 4(a)). In order to assess the error introduced by processing and diluting high-turbidity samples, we subdivided the samples into categories greater or less than the highest turbidity standard (800 NTU). When including only samples with turbidity <800 NTU, the correlation between turbidity and TSS improved to an r^2 value of 0.92 (TSS = 1.1 × turbidity + 4.8) (Fig. 4(b)). For samples with turbidity >800 NTU (which were all collected during the storm events of 7, 13 and 14 July, and 7 September 2011), the correlation decreased to an r^2 value of 0.769 (TSS = 1.5 × turbidity + 1300). These results suggest that measurement of turbidity, a much easier parameter to measure than TSS, may be used to estimate TSS and hence sediment loading in the study watershed. In-stream turbidimeters, however, may not accurately record the largest sediment loads in this watershed because these instruments have saturation limits that may be exceeded during storm events (Anderson, 2005).

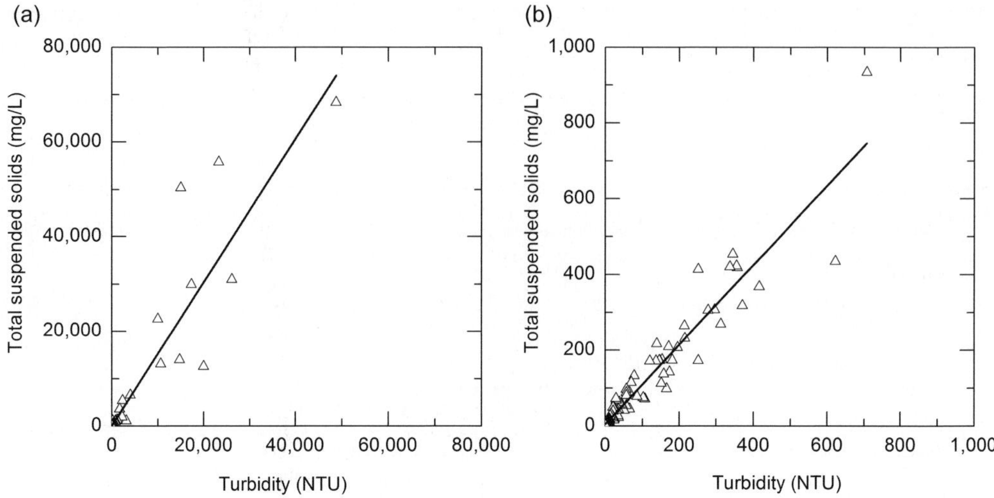

Fig. 4 Relation of TSS and turbidity for: (a) all samples and (b) samples with a turbidity < 800 NTU.

CONCLUSION

During periods of no precipitation, including during snowmelt runoff, turbidity and TSS values at sites upstream and downstream from a burned area in the Colorado Front Range were low and typically similar among sites. Low-intensity storm events often led to increases in turbidity and TSS at all sites, but the increases were greater at sites within or downstream from the burned area. High-intensity convective storms resulted in up to 80-fold increases in stream discharge downstream from the burned area, whereas discharge upstream from the burned area increased only 0.4-fold. These high-intensity convective storms mobilized sediment from hillslopes and led to substantial increases in turbidity and TSS in the local streams. The re-working of sediment deposits within the stream channel commonly resulted in elevated turbidity levels for weeks after the high-intensity convective storms. Subsequent, low-intensity rain events remobilized additional stream sediment. The results from this study suggest that dramatic water quality changes can occur during and immediately after high-intensity storms. A network of high-density precipitation stations and stream gauges is necessary to understand more fully water quality and quantity issues related to such events in a landscape impacted by wildfire.

Acknowledgements Many people assisted in the preparation of this paper by providing field or laboratory assistance, information, or review. We thank Kenna Butler, Jennifer Carter, Brian Ebel, Deborah Martin, John Moody, Deborah Repert, and Doug Winter (USGS), Jennifer Morse (University of Colorado), the USGS Colorado Water Science Center, the City of Boulder, Boulder County Parks & Open Space, Urban Drainage and Flood Control Disrict, Pine Brook Hills Water District, and several private property owners. Support was provided by the US Geological Survey and by National Science Foundation Grant #0724960 (the Boulder Creek Critical Zone Observatory). Any use of trade, product, or firm names is for descriptive purposes only and does not imply endorsement by the US Government.

REFERENCES

Anderson, C. W. (2005) Turbidity. In: *US Geological Survey Techniques of Water-Resources Investigations, Book 9, Chap. A6.7*, accessed March 13, 2012, at http://pubs.water.usgs.gov/twri9A.

Benavides-Solorio, J. & MacDonald, L. H. (2005) Measurement and prediction of post-fire erosion at the hillslope scale, Colorado Front Range. *Int. J. Wildland Fire* 14, 1–18.

Beschta, R. L. (1990) Effects of fire on water quantity and quality. *In* Natural and Prescribed Fire in Pacific Northwest Forests (ed. by J. D. Walstad, S. R. Radosevich & D. V. Sandberg). Oregon State University Press, Corvallis, Oregon, 219–232.

Boulder County (2012) Fourmile Canyon Fire Flood Mitigation and Land Rehabilitation: accessed 19 January 2012, at http://www.bouldercounty.org/live/environment/land/pages/4milefirelandrehab.aspx.

Fourmile Emergency Stabilization Team (2010) Fourmile Emergency Stabilization Burned Area Report, Boulder County, Colorado, 14 p.

Gresswell, R. E. (1999) Fire and aquatic ecosystems in forested biomes of North America. *Trans. Am. Fisheries Soc.* 128, 193–221.

Hach Company (2010) *Basic user manual, 2100Q and 2100Qis*, Edition 1: Accessed March 13, 2012, at www.hach.com.

Holden, Z. A., Morgan, P., Crimmins, M. A., Steinhorst, R. K. & Smith, A. M. S. (2007) Fire season precipitation variability influences fire extent and severity in a large southwestern wilderness area, United States. *Geophys. Res. Lett.* 34, L16708.

Keeley, J. E. (2009) Fire intensity, fire severity, and burn severity: a brief review and suggested usage. *Int. J. Wildland Fire* 18, 116–126.

Kirchner, J. W., Finkel, R. C., Riebe, C. S., Granger, D. E., Clayton, J. L., King, J. G. & Megahan, W. F. (2001) Mountain erosion over 10 yr, 10 k.y., and 10 m.y. time scales. *Geology* 29, 591–594.

Kirchner, J. W., Feng, X., Neal, C. & Robson, A. J. (2004) The fine structure of water-quality dynamics: the (high-frequency) wave of the future. *Hydrol. Processes* 18, 1353–1359.

McCleskey, R. B., Writer, J. H. & Murphy, S. F. (2012) Water chemistry data for surface waters impacted by the Fourmile Canyon wildfire, Colorado, 2010–2011. *US Geological Survey Open-File Report* (in press).

Miller, J. F., Frederick, R. H. & Tracey, R. S. (1973) *NOAA ATLAS 2, Precipitation-Frequency Atlas of the Western United States*. US Dept. of Commerce, NOAA, National Weather Service, Washington, DC, USA.

Moody, J. A. & Martin, D. A. (2001) Initial hydrologic and geomorphic response following a wildfire in the Colorado Front Range. *Earth Surface Processes and Landforms* 26, 1049–1070.

National Atmospheric Deposition Program (2011) NADP Maps and Data: Accessed 23 November 2011 at http://nadp.sws.uiuc.edu/.

Neary, D. G., Ryan, K. C. & DeBano, L. F. (eds) (2005) Wildland fire in ecosystems: effects of fire on soils and water: USDA General Technical Report RMRS-GTR-42-vol.4. Ogden, Utah, USDA Forest Service Rocky Mountain Research Station. 250 p.

Rantz, S. E., *et al.* (1982) Measurement and computation of streamflow. *US Geological Survey Water-Supply Paper 2175*, 631 p.

Robichaud, P. R., Ashmun, L. E. & Sims, B. D. (2010) Post-fire treatment effectiveness for hillslope stabilization: USDA General Technical Report RMRS-GTR-240. Fort Collins, Colorado, USDA Forest Service Rocky Mountain Research Station. 62 p.

Ruddy, B. C., Stevens, M. R., Verdin, K. L. & Elliott, J. G. (2010) Probability and volume of potential post-wildfire debris flows in the 2010 Fourmile burn area, Boulder County, Colorado. *US Geological Survey Open-file Report 2010–1244.*

Smith, H. G., Sheridan, G. J., Lane, P. N. J., Nyman, P. & Haydon, S. (2011) Wildfire effects on water quality in forest catchments: a review with implications for water supply. *J. Hydrol.* 396, 170–192.

Tiedemann, A. R., Conrad, C. E., Dieterich, J. H., Hornbeck, J. W., Megahan, W. F., Viereck, L. A. & Wade, D. D. (1979) Effects of fire on water: a state-of-knowledge review. *USDA Forest Service General Technical Report WO-10*, Washington, DC, 28 p.

Urban Drainage and Flood Control District (2011) ALERT system: accessed 23 November 2011 at http://alert.udfcd.org/datadisp.html.

US Geological Survey (2011) National Water Information System (NWISWeb): US Geological Survey database accessed 15 December 2011 at http://waterdata.usgs.gov.

Westerling, A. L., Hidalgo, H. G., Craven, D. R. & Swetnam, T. W. (2006) Warming and earlier spring increase western U.S. forest wildfire activity. *Science* 313, 940–943.

Wondzell, S. M. & King, J. G. (2003) Post-fire erosional processes in the Pacific Northwest and Rocky Mountain regions. *Forest Ecology and Management* 178, 75–87.

Writer, J. H., McCleskey, R. B. & Murphy, S. F. (2012) Effects of wildfire on source-water quality and aquatic ecosystems, Colorado Front Range. In: *Wildfire and Water Quality: Processes, Impacts and Challenges* (Proc. conference held in Banff, Canada, June 2012). IAHS Publ. 354, IAHS Press. Wallingford, UK (this volume).

Fire and sediment in an upland stream in Hong Kong

MERVYN. R. PEART[1], LINCOLN FOK[2] & RONALD. D. HILL[3]

1 *Department of Geography, University of Hong Kong, Hong Kong*
mrpeart@hku.hk
2 *Department of Science and Environmental Studies, Hong Kong Institute of Education, Hong Kong*
3 *Department of Ecology and Biodiversity, University of Hong Kong, Hong Kong*

Abstract Hill fires are common in Hong Kong. In February 2007, a hill fire burnt an area of around 1 km^2 on the hillslopes of Tai To Yan providing the opportunity to examine the effects of fire. Descriptive statistics in the form of the 25th, 50th and 75th percentiles show an increase in storm-period suspended sediment concentration compared to the year preceding the fire (2006). Similar results were observed for a fire in December 2004. Storm-period suspended sediment data for the years 2008–2011 show evidence of a return to pre-fire levels. Regular weekly sampling for suspended sediment for the years 2006 to 2010 indicates that fire resulted in an increase in the dry season median value in 2007 compared to that in 2006, the year preceding the burn. Weekly sampling has been undertaken during 2006 to 2010 for chlorophyll-a in the stream water, and the results are elaborated upon in terms of the impact of fire on water quality.

Key words fire; suspended sediment; chlorophyll-*a*; statistical comparison

INTRODUCTION

In Hong Kong, wildfire is referred to as hill fire. Marafa & Chau (1999) have observed that half of the area within Hong Kong country-park boundaries (442 km^2) has been burnt at least once in the last 12 years, whilst Owen & Shaw (2007) suggest that about 5% of the land area is burnt annually. Over a 10-year period from 2001/2 to 2009/10, the Agriculture, Fisheries and Conservation Department reported 551 hill fires within or threatening Hong Kong Country Parks, whilst the Fire Services Department attended 1295 and 819 vegetation fires in 2009 and 2010, respectively. The existence of grassland and shrubland on the hillslopes of Hong Kong provides the fuel for wildfires, and these are the types of vegetation most commonly affected (Chan, 2005). Chan (2005) also noted that hill fires tended to be more frequent on slopes with an easterly to southerly aspect, which gave a longer exposure to sunlight thereby promoting drier fuel, often grass, for ignition. The study by Chan (2005) revealed that the elevations with the highest fire frequency were 190–380 m PD. Generally speaking, hill fire is not natural in Hong Kong (Chan, 2005) and most fires result from human activity. Both Chau & Corlett (1994) and Chan (2005) attest to the seasonality of hill fire occurrence in Hong Kong; they are typically associated with the dry season (October to April). The high rainfall and higher humidity levels in summer do not promote the ignition and spread of hill fires, whilst Chan (2005) suggests that the comparatively low relative humidity in the dry season may be an important control. The two hill fires that were included in this study occurred in December 2004 and January 2007, coinciding with the dry season.

Hill fires have been demonstrated to affect soil erodibility in Hong Kong (Ternan & Neller, 1999) whilst Vohora & Donoghue (2004) indicate a causal linkage to slope degradation, including landslides. Peart *et al.* (2009) have identified that hill fire impacts upon soil erosion and suspended sediment transport in Hong Kong. A recent hill fire in early February 2007, in a small upland catchment, of which 80% burned, afforded a further opportunity to examine the effects of fire upon sediment production in Hong Kong and the observations are presented in this paper.

STUDY AREA AND METHODS

Hong Kong has a subtropical climate with a hot, wet summer and cool, dry winter. Climatological data for 1971–2000 give an annual average rainfall of 2383 mm, of which more than 80% occurred in the months of April to September. Mean annual air temperature is 23.1°C, with

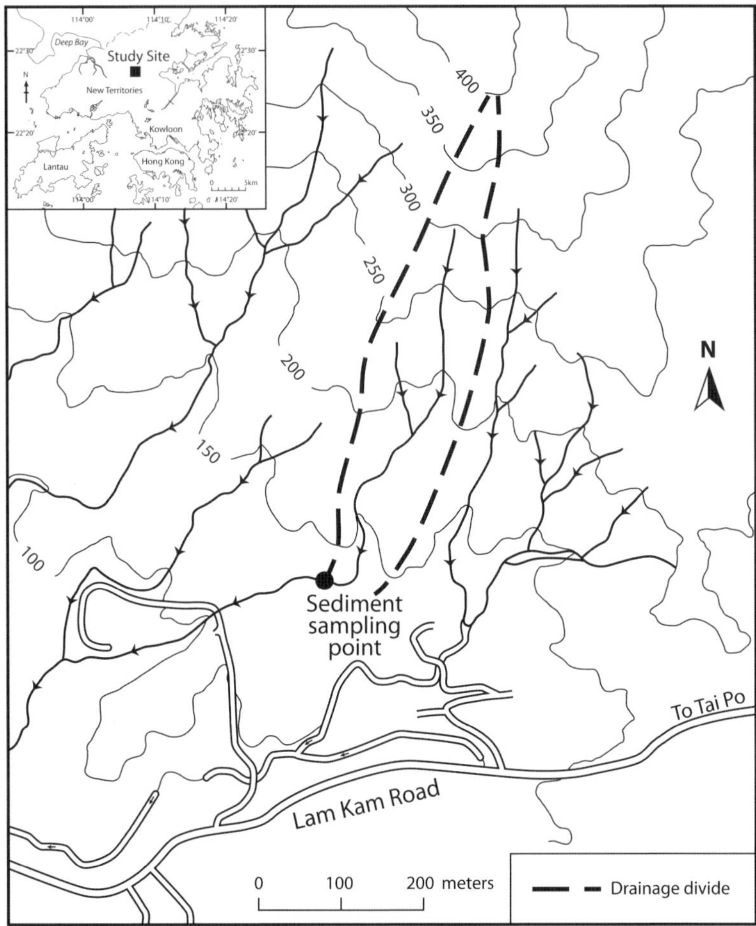

Fig. 1 Location of the study area.

monthly mean air temperatures ranging from 16.1°C to 28.7°C. Mean annual relative humidity for 1971–2000 is 78% and ranges from 69% in December to 84% in May. The topography of Hong Kong may be described as rugged, with altitudes ranging up to 957 m PD on the highest peak of Tai Mo Shan. Where the soil is poor or frequently affected by fire hillslopes may be covered by grass, but in other areas shrubs may dominate. Some woodland and plantations are observed on hillslopes. About 40% of the land area is designated under the Country Parks Ordinance.

The study area (Fig. 1), located on the slopes of Tai To Yan in the Northwest New Territories, has a vegetation cover predominantly comprising grass, fern and shrubland (Hill *et al.*, 2004), which is typical of Hong Kong (Dudgeon & Corlett, 2004). Fires have occurred in the area with burns recorded in 1988, 1990, 1992, 1995, 1998, 2004 and 2007, and Chan (2005) also confirms that the area is susceptible to fire. The occurrence of fire has influenced vegetation development in the area and is reflected in the dominance of grass, fern and low shrubland. The geology consists of volcanic tuffs with some siltstone/sandstone interbeds and granodiorite. Colluvium is widespread on the lower slopes. Small-scale landsliding occurred in the upper source area of the drainage system in 2001 and 2003, triggered by rainfall. For the period 1999–2010, mean annual rainfall at the nearby Kadoorie Farm and Botanic Garden gave an average annual rainfall of 2772.9 mm, with considerable inter-annual variation.

Suspended sediment production has been monitored in this small upland headwater catchment (0.052 km^2), which is located in the Kam Tin drainage basin. The outlet of this small basin is around 120 m PD, whilst the highest point is around 380 m PD. The average slope gradient is 30°, with some areas being over 50°, which compares with between 15 and 60° for 66% of Hong Kong slopes (Styles & Hansen, 1989). Suspended sediment sampling began in this drainage basin from 1993, with typical sample volumes being 400–500 ml. Manual sampling, from the centre of the <1-m wide channel, and an ISCO model 2700 automatic sampler, were used to collect samples under a range of streamflow conditions and stage-levels throughout any given year. The sampling strategy has included periods of regular-interval weekly sampling. Upon return to the laboratory, the suspended sediment was separated by filtration using pre-weighed GF/C filter papers and suspended sediment concentration determined by oven drying (at 103–105°C) to a constant weight. For determination of chlorophyll-a, sampling was undertaken on a regular weekly schedule. Water samples were collected, transferred to the laboratory and filtered as soon as possible through a GF/C filter paper. Acetone was used to extract the chlorophyll and a spectrophotometer measured extinction at 665 nm. The method is based upon Golterman (1970) and Lorenzen (1967). For both suspended sediment and chlorophyll-a, the non-parametric Wilcoxon rank-sum test, which avoids the assumptions regarding data associated with parametric tests, has been used to compare the data and the 95% significance level adopted. The tests were done using the R-project statistical package.

RESULTS

The summary descriptive statistics of the storm period suspended sediment concentration data (2004–2011) for the study basin are presented in Table 1. From Table 1, it can be seen that the two years in which the basin was affected by fire, namely 2005 and 2007, exhibited an increase in suspended sediment for the percentile values compared to the preceding and non-fire years of 2004 and 2006, respectively. Moreover, there is a statistical difference at the 95% significance level between 2004 *vs* 2005 and 2006 *vs* 2007, suggesting that fire raised suspended sediment in the stream compared to the preceding years. The respective increases in medians are 8.3 and 3.0 mg/L, which equate to 148% and 16%. Comparison of 2007 with 2008 median values in Table 1 reveals that 2008 is higher compared to 2007 by 7.3 mg/L, and that there is no statistically significant difference between the two data sets at the 95% level, suggesting that suspended sediment levels remained high in 2008 over 12 months after the fire of January 2007. Aggregation of the years 2008, 2009, 2010 and 2011 gives a median value of 16.8 mg/L, lower than the 21.7 mg/L observed

Table 1 Storm period suspended sediment statistics (mg/L).

Year	Percentiles: 25	50	75	Sample size
2004	2.7	5.6	10	77
2005*	5.7	13.9	34.5	196
2006	9.8	18.7	41.8	229
2007*	11.3	21.7	61.8	173
2008	12.5	29	67.5	128
2009	3.7	8.9	18.3	164
2010	16.7	36	92.4	73
2011	5.7	12.9	30.1	89
2005 + 2007*	8.5	17.8	46.3	369
2004 + 2006	6.6	13.8	37.2	306
2006 + 2008	10.2	22.6	50.2	357
2008, 2009, 2010 + 2011	6.5	16.8	43.7	454

* fire affected

Table 2 Weekly sampling median suspended sediment concentrations (mg/L).

Year	Dry season	Sample size	Wet season	Sample size
2006	3.52	20	5.8	17
2007*	5.74	17**	5.51	21
2008	3.17	22	5.63	24
2009	4.9	18	3.15	18
2010	4.04	25	8.3	24
2009–2010	4.49	106	5.69	104

* fire affected
** no data in January pre-fire

Table 3 Weekly sampling median chlorophyll-a concentrations (μg/L).

Year	Dry season	Sample size	Wet season	Sample size
2006	0.33	20	0.28	17
2007*	0.45	17**	0.38	21
2008	0.19	22	0.33	24
2009	0.16	18	0.22	18
2010	0.23	25	0.22	24
2006–2010	0.24	106	0.24	104

* fire affected
** no data in January pre-fire

for the 2007 fire-affected year. Statistical testing of the 2007 year against the combined 2008–2011 data sets reveals a significant difference at the 95% level. This suggests that suspended sediment concentrations are gradually declining to below post-2007 levels.

Regular weekly sampling of suspended sediment has been undertaken in the drainage basin, and at the same time bulk water samples were collected for chlorophyll-a determination. The results from the regular sampling programme for suspended sediment and chlorophyll-a are presented in Tables 2 and 3, respectively. In terms of suspended sediment, the annual median suspended sediment concentrations for the years 2006–2010 range from 3.17 to 5.74, and 3.15 to 8.30 mg/L, respectively, for the dry and wet seasons. The fire year of 2007 does have the highest dry season median value, and at 5.74 mg/L the concentration is 2.2 mg/L, or 63%, higher than the 2006 value of 3.52 mg/L. The fire impacted dry season data for 2007 are significantly different at the 95% level from the dry season data of 2006, 2009 and 2010, but not 2008. It should be noted that the concentrations are very low and are also much smaller than for the equivalent storm period median values during corresponding years. Comparison of the wet season data for the regular interval sampling is complicated by the occurrence of storm events.

Regarding chlorophyll-a, Table 3 reveals that the dry season samples for 2007 exhibited the highest median, with a value of 0.45 μg/L. However, the variation across the years is small. Statistical testing reveals no significant difference at the 95% level between 2006 *vs* 2007, and 2007 *vs* 2006, 2008, 2009 and 2010. The wet season data also reveals limited inter-annual variation in chlorophyll-a, and there is no statistically significant difference at the 95% significance level between 2007 and other years. In general, the concentrations for chlorophyll-a are low and for the 2006–2010 data, the median dry and wet season chlorophyll-a values are 0.24 and 0.24 μg/L, respectively. Regarding chlorophyll-a, the fire of January 2007 did therefore not result in great change on the basis of the data assembled using regular weekly sampling.

DISCUSSION

As noted above, the data in Table 1 suggest that in the year immediately following fires, namely 2005 and 2007, the median storm period suspended sediment concentration was higher than in the preceding year: hill fire results in an increase in storm period sediment concentrations. Data for two other fire years in the same basin, namely 1996 and 1998, show the same pattern with median storm-period suspended sediment concentrations being 5.3 and 2.3 mg/L higher, respectively, than in the year preceding the fire; these are equivalent to 88% and 24% increases. The two hill fire events of December 2004 and February 2007 provide evidence that the burning of vegetation can result in enhanced suspended sediment concentrations. An increase in suspended sediment concentrations following fire has been reported by previous work, including, amongst others, Veenhuis & Bowman (2002) and Malmon *et al.* (2007). Table 1 shows that the annual median suspended sediment concentrations for the years 2006 and 2008 were greater than the fire impacted years of 2005 and 2007, respectively. This suggests that the impact from the fires of December 2004 and February 2007 on the storm-period suspended sediment may extend beyond the first 12 months following a burn. This finding is in contrast to data reported by Peart *et al.* (2009) for earlier fire events in the study drainage basin. The prolongation of the effects of fire upon suspended sediment concentrations may be due to climate, as 2006 and 2008 were comparatively wet with annual rainfall totals at the adjacent Kadoorie Farm and Botanic Gardens of 3256.5 mm and 3483.7 mm, respectively. In addition, Kunze & Stednick (2006) have shown that sediment storage can affect post-fire sediment production. It should also be noted that there is a statistically-significant difference between 2007 and the combined 2008, 2009, 2010 and 2011 data sets indicating a decline in storm-period sediment concentrations over the longer term after the February 2007 fire. Bearing in mind any limitations of the sampling strategy, evidence of a change in sediment concentrations following a hill fire in an upland catchment has been presented. However, it should be noted that the response of suspended sediment to fire may be complicated by, for example, sediment delivery and storage (Kunz & Stednick, 2006; Shakesby & Doerr, 2006), the type and sequence of storm events, both pre- and post- fire (Desilets *et al.*, 2007; Smith *et al.*, 2010) and, the extent and severity of the burn (Rhoades *et al.*, 2011).

Evidence of an increase in storm period suspended sediment concentrations following hill fire has been presented. However, the median values and 75th percentile values presented in Table 1 show that the concentrations are comparatively low. This may reflect vegetative cover regeneration. For example, for the fire of December 2004, which affected sediment production in 2005, vegetation cover on three monitoring plots had reached 88%, 79% and 70% by November 2005 and, at one other plot, 99% by August 2005. For the fire of February 2007, data from two monitoring plots reveal live vegetation coverage of 25% and 33% by 11 April 2007, 65% and 71% by 24 May 2007, and 98% and 97% by 28 August 2007. A further contributing factor may be that during the hill fire of January 2007, soil moisture associated with the drainage system prevented ignition and burning of vegetation in some of the riparian zone. This riparian vegetation may have acted as a filter, reducing sediment delivery to the basin outlet.

The study basin stream is part of the Kam Tin drainage basin, which is one of the largest in Hong Kong. As part of the Hong Kong Governments attempts to improve water quality in surface waters, a series of Water Quality Objectives have been introduced. For the Kam Tin drainage basin, the suspended sediment Water Quality Objective is a median value of <20 mg/L based upon monthly sampling. Table 1 reveals that for the eight years of data, only three (2007, 2008 and 2010) have median values that exceeded 20 mg/L. Moreover, for the eight years of data presented in Table 1, the highest 75th percentile was only 92.4 mg/L, which is low. For example, in comparison Peart (1997) reports a storm period sampling 75th percentile value of 954 mg/L for 346 samples, for the period 1991–1995 in the nearby Lam Tsuen River, which during 1991 and 1992 was impacted by roadwork's. The low values for the 75th percentiles observed in this study basin are evidence of the protective effect of vegetation that has previously been demonstrated by Lam (1978) in a study of three small catchments at Tai Lam Chung, also in the New Territories of Hong Kong.

The concentrations of chlorophyll-a in streams and rivers may be influenced by physical, chemical and biological factors and processes (Neal *et al.*, 2006). If the occurrence of fire in the basin had resulted in a large change in any of the factors controlling chlorophyll-a concentrations, then it might be expected that 2007 would have distinctive levels of chlorophyll-a. For example, the burn of 2007 could have reduced the shade afforded to the stream channel by riparian vegetation allowing more sunlight and thereby stimulating primary production of aquatic algae. Light availability is an important driver of primary production in tropical streams and rivers and green algae require much higher light intensities compared to diatoms or cyanobacteria (Davies *et al.*, 2008). Indeed, Dudgeon (1988) has shown that for Hong Kong streams, the lowest standing stock of periphyton occurred in shaded streams. A further control on light availability is suspended matter (Lewis, 2008). Table 2 illustrates that regular weekly sampling reveals no great difference in suspended sediment concentrations for the year affected by the hill fire, 2007, compared to the years of 2006, 2008, 2009, 2010 and 2011 and, furthermore, Table 1 indicates that whilst fire may result in higher median values of storm-period suspended sediment compared to the preceding year, the increases were 8.3 and 3.0 mg/L, respectively, for the fire years of 2005 and 2007. These values are too low to result in a change in light availability (Lewis, 2008). Ganf & Rea (2007) reported that the Queensland Water Quality Guidelines for the wet tropics recommend that chlorophyll-a concentrations for rivers should be <3 μg/L. The annual median values presented in Table 3 are well below this value and for the years 2006, 2007, 2008, 2009 and 2010, only zero, four, three, zero and two recorded values were greater than 3 μg /L, respectively.

In terms of comparison, Zhang *et al.* (2011) reported values for chlorophyll-a in the Pearl River Estuary of from around 0.83 to 11.77 μg/L, with the observed values in the study catchment being similar to the lower values reported by Zhang *et al.* (2011). Table 3 suggests that fire has not impacted the chlorophyll-a content of water in the study stream as 2007 is similar to the other years for which chlorophyll-a data are available.

CONCLUSION

Based upon observations in a small upland drainage basin in Hong Kong and bearing in mind any limitations of the sampling strategy, hill fire may result in an increase in storm-period suspended sediment concentrations. Data for 2008, 2009, 2010 and 2011 suggest a gradual decline of storm-period sediment concentrations to pre-fire levels. Regular weekly sampling suggests an increase in dry season sediment concentrations (which reflect low-flow conditions) in the fire affected year of 2007, compared to the dry season of the preceding year, but the suspended sediment concentrations are in general low. Chlorophyll-a values in runoff from the basin appear not to have changed post-fire.

Acknowledgements Funding from the Hui-Oi-Chow Trust Fund is gratefully acknowledged.

REFERENCES

Chan, Wu-wah (2005) A feasibility study of hill fire management in Hong Kong Country Parks using GIS analysis. MSc Dissertation, Environmental Management, University of Hong Kong, 85 pp.

Chau, K. C. L. & Corlett, R. T. (1994) Fire and weather in Hong Kong. In: *Proceedings 12th Conference on Fire and Forest Meteorology* (October 1993, Georgia, USA), 442–452. Society of American Foresters.

Davies, P. M., Bunn, S. E. & Hamilton, K. (2008) Primary production in tropical streams and rivers. In: *Tropical Stream Ecology* (ed. by D. Dudgeon), 23–42. Elsevier, Amsterdam.

Desilets, S. L. E., Nijssen, B., Ekwurzel, B. & Ferré, Ty, P. A. (2007) Post-wildfire changes in suspended sediment rating curves: Sabino Canyon, Arizona. *Hydrol. Processes* 21, 1413–1423.

Dudgeon, D. (1988) The influence of riparian vegetation on macro invertebrate community structure in four Hong Kong streams. *J. Zool. Lond.* 216, 609–627.

Dudgeon, D. & Corlett, R. (2004) *The Ecology and Biodiversity of Hong Kong*. Friends of the Country Parks and Joint Publishing, Hong Kong, 336 pp.

Ganf, G. G. & Rea, N. (2007) Potential for algal blooms in tropical rivers of the Northern Territory, Australia. *Marine and Freshwater Research* 58, 315–326.

Golterman, H. L. (1970) *Methods for Analysis of Fresh Waters*. IBP Handbook no. 8. Blackwell Scientific, Oxford.

Hill, R. D., Peart, M. R. & Guan, D.-S, (2004) The effects of annual harvesting on the subsequent phytomass and species composition of grassland and fernland: a Hong Kong case, Singapore *J. Tropical Geogr.* 25(1), 77–91.

Kunz, M. D. & Stednick, J. D. (2006) Streamflow and suspended sediment yield following the 2000 Bobcat fire, Colorado. *Hydrol. Processes* 20, 1661–1681.

Lewis, Jr, W. M. (2008) Physical and chemical features of tropical flowing waters. In: *Tropical Stream Ecology* (ed. by D. Dudgeon), 1–21. Elsevier, Amsterdam.

Lam, K. C. (1978) Soil erosion, suspended sediment and solute production in three Hong Kong catchments. *J. Tropical Geog.* 47, 51–62.

Lorenzen, C. J. (1967) Determination of chlorophyll and pheo-pigments: spectrophotometric equations. *Limn. and Oceanogr.* 12, 342–346.

Malmon, D. V., Reneau, S. L., Katzman, D., Lavine, A & Lyman, J. (2007) Suspended sediment transport in an ephemeral stream following wildfire. *J. Geophys. Res.* 112. doi: 0.1029/2005 FJ000459, 2007.

Marafa, L. M. & Chow, K. C. (1999) Effect of hill fire on upland soil in Hong Kong. *Forest Ecology & Manage.* 120, 97–104.

Neal, C., Hilton, J., Wade, A. J., Neal, M. & Wickham, H. (2006) Chlorophyll-a in the rivers of eastern England. *Sci. Total Environ.* 365, 84–104.

Owen, B. & Shaw (2007) *Hong Kong Landscapes: Shaping the Barren Rock*. Hong Kong University Press. 253 pp.

Peart. M. R. (1998) Human impact upon sediment in rivers: some examples from Hong Kong. In: *Human Impact on Erosion and Sedimentation* (ed. by D. E. Walling & J.-L. Probst), 111–118. IAHS Publ. 245. IAHS Press, Wallingford, UK.

Peart, M. R., Hill, R. D. & Fok, L. (2009) Environmental change, hillslope erosion and suspended sediment: some observations from Hong Kong. *Catena* 79, 198–204.

Rhoades, C. C., Entwistle, D. & Butler, D. (2011) The influence of wildfire extent and severity on streamwater chemistry, sediment and temperature following the Hayman fire, Colorado. *Int. J. Wildland Fire* 20, 430–442.

Shakesby, R. A. & Doerr, S. H. (2006) Wildfire as a hydrological and geomorphological agent. *Earth-Science Revs* 74, 269–307.

Smith, H. G., Sheridan, G. J., Lane, P. N. J. & Sherwin, C. B. (2010) Paired *Eucalyptus* forest catchment study of prescribed fire effects on suspended sediment and nutrient exports in south-eastern Australia. *Int. J. Wildland Fire* 19(5), 624–636.

Styles, K. A. & Hansen, A. (1989) *Geotechnical Areas Studies Programme Territory of Hong Kong*. Geotechnical Control Office, Civil Engineering Services Department. Hong Kong GASP Report XII. 345 pp.

Ternan, J. L. & Neller, R. (1999) The erodibility of soils beneath wildfire prone grasslands in the humid tropics, Hong Kong. *Catena* 36, 49–64.

Veenhuis, J. E. & Bowman, P. R. (2002) Effects of wildfire on the hydrology of Frijoles and Capulin canyons in and near Bandelier National Monument, New Mexico. *USGS Fact Sheet, 141-02. 4P.*

Vohora, U. K. & Donoghue, S. L. (2004) Application of remote sensing data to landslide mapping in Hong Kong. In: *Geo-Imagery Bridging Continents* (Proc. ISPRS Congress, July 2004 Istanbul, Turkey), vol. XXXV, part B5, 489–493.

Yuanzhi Zhang, Hui Lin, Chuqun Chen, Liding Chen, Bing Zhang & Gitelson, A. A. (2011) Estimation of Chlorophyll-a concentration in estuarine waters: case study of the Pearl River estuary, South China Sea. *Environ. Res.* 6(2), doi:10.1.1088/1748-9326/6/2/024016.

Changes in benthic community structure and function in an Australian regulated upland stream following wildfire

MICHAEL A. REID & MARTIN C. THOMS
Riverine Landscapes Research Laboratory, University of New England, Armidale, NSW 2351, Australia
mreid24@une.edu.au

Abstract The effects of fire on stream ecosystems are mostly indirect and can be attributed to post-fire floods, enhanced sediment and nutrient influxes, and channel morphology changes. Flow regulation will modify water and sediment regimes and thus can be expected to influence the nature and trajectory of post-fire changes in regulated streams. Despite this, few studies have examined how fire-related disturbances interact with flow regulation. This study draws on the results of several previous projects to generate multiple lines of evidence to explore pre- and post-fire benthic metabolism and benthic macroinvertebrate assemblages in an Australian regulated upland stream affected by wildfire during the austral summer of 2003. The benthic metabolism results show increased autotrophy following the fire, probably due to higher nutrient influxes and light availability because of reduced shading in conjunction with reduced respiration due to changes in the quality of carbon inputs and reduced interaction with the hyporheos associated with bed armouring.

Key words macroinvertebrates; benthic metabolism; flow regulation; wildfire

INTRODUCTION

The direct effects of wildfire on stream communities are generally small (Minshall 2003; Vieira *et al.*, 2004; Malison & Baxter, 2010); however, indirect effects can be both substantial and persistent. Indirect effects include hydrological disturbances and channel changes because of the effect of fire on catchment runoff rates and volumes (Minshall, 2003), increases in the delivery of sediment to streams as a result of the loss of ground cover vegetation (Silins *et al.*, 2009; Smith *et al.*, 2011a,b), changes in water and sediment quality (Smith *et al.*, 2010,a,b) and increased primary productivity as a result of reduced shading and higher nutrient fluxes from catchment soils (Fernandez *et al.*, 2011; Rhoades *et al.*, 2011). The effect of wildfires may be evident decades after the initial event (Robinson *et al.*, 2005); nevertheless, several studies have shown that instream aquatic communities can recover substantial structural and functional elements of their pre-fire state within a few years of burning. Benthic communities are thus considered to be resilient to the effects of catchment wildfire (Minshall, 2003; Peat *et al.*, 2005).

Recovery of stream communities following wildfires is hampered in streams whose catchments are subject to other anthropogenic disturbances such as logging, livestock grazing and fragmentation associated with water resource development (Minshall, 2003; Beschta *et al.*, 2004; Mellon *et al.*, 2008). Importantly, despite the evidence that interactions with anthropogenic disturbances strongly influence recovery rates and even trajectories, relatively few studies have examined recovery in streams regulated by dams. Flow regulation directly affects sediment and water regimes, both key drivers of stream disturbance following fire. Moreover, given the important role that longitudinal connectivity is likely to play in facilitating recolonisation of severely disturbed reaches, fragmentation by dams and weirs has the potential to influence further both recovery rates and trajectories.

Here we draw together the results of previous studies of benthic macroinvertebrate communities and benthic metabolism in a regulated, middle-order gravel-bed stream affected by a severe and extensive wildfire with the aim of extending the times series of data on the recovery trajectory of the stream following fire. The original studies from which we draw these data were conducted primarily to investigate questions other than post-fire recovery and, as such, the study designs and available data limit the strength of the inferences that can be made. Nevertheless, the data provide a valuable window into the ways in which flow regulation may modify recovery rates and trajectories in forested stream ecosystems impacted by the occurrence of wildfire.

METHODS

Study area

The Cotter River is an upland cobble and gravel-bed system that drains a catchment of approximately 483 km^2 located within the Brindabella Ranges in the Australian Capital Territory (Fig. 1). Although three dams regulate flow, the catchment itself is largely unmodified by human land use. These dams supply water for the city of Canberra (pop. ~322 000); however, environmental flow releases designed to minimize the impact of the dams and mimic the natural flow regime are made. Sites included in the study are all situated on the reach of the Cotter River between the Bendora and Cotter dams (Fig. 1) at altitudes ranging from ~700 to 500 m ASL. The climate is temperate with hot summers and cold winters. Average precipitation ranges from 990 to 1080 mm, with the wettest months between July and October. During the summer of 2003, the Cotter River catchment was subject to severe and extensive fire. These fires burnt a total of 260 000 ha in the ACT and NSW, including most of the Cotter River catchment above the Cotter Dam (McRae, 2003).

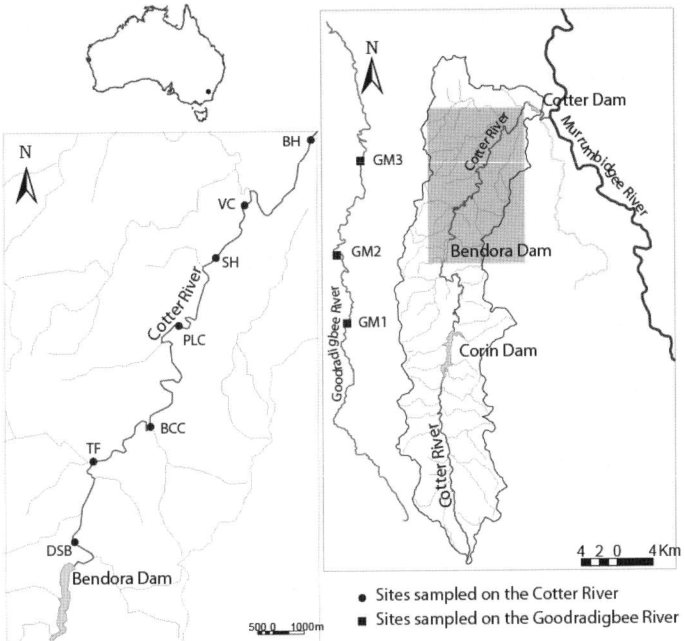

Fig. 1 The study area showing the locations of the Cotter and Goodradigbee rivers and the river locations sampled in Chester (2003), Norris & Thoms (2004), Peat *et al.* (2005), Reid *et al.* (2006) and Reid & Thoms (2008), which are included in this overview.

Study design

This study incorporates data from several previous projects on benthic macroinvertebrates and benthic metabolism collected from the Cotter River (Fig. 1) and from a nearby unregulated stream. The details of the original source data including timing of sampling and the variables collected are summarised in Table 1. The methods used in each case are described in the original publications (Chester, 2003; Peat *et al.*, 2005; Reid *et al.*, 2006; Reid & Thoms, 2008). Comparisons between variables derived from different studies are complicated by the different methods used in the original studies, which reflects the different purposes for which the data were collected in each case. For this reason, variables chosen for comparison were selected to minimise the possible confounding effects of sampling methods. Accordingly, benthic metabolism is compared only

Table 1 Details (variables, sample dates, locations and rivers) of source data used in this study.

Variable		P/R (productivity/respiration ratio)								Macroinvertebrate sampling								
River		Cotter						Goodradigbee		Cotter						Goodradigbee		
Site		VC	TF	SH	PLC	DSB*	BH	GM2	GM1	VC	TF	PLC	DSB*	BH	BCC	GM3	GM2	GM1
	Date																	
Chester (2003)	Feb 2002					×												
	May 2002					X												
Norris & Thoms (2004)	Apr 2003				X	X	X		X									
	May 2003				X	X	X		X									
	Jun 2003									X	X	X	X	X				
	Jul 2003				X		X	X	X									
	Aug 2003				X	X	X	X	X									
	Oct 2003				X	X	X		X									
	Nov 2003											X						
Peat *et al.* (2005)	Oct 2001												X			X	X	X
	Oct 2003												X			X	X	X
	Oct 2004												X			X	X	X
Reid *et al.* (2006)	Nov 2003		X		X													
	Dec 2003		X	X	X													
	Mar 2004		X	X	X													
	May 2004		X	X	X													
Reid & Thoms (2008)	Feb 2006									X	X	X			X			

*referred to as CM2 in Chester (2003) and *Peat et al.* (2005) and as 4102005 in Norris & Thoms (2004).

using productivity/respiration (P/R) ratios rather than primary productivity and/or respiration rates. In the case of benthic macroinvertebrates, sampling in the different studies varied somewhat: Norris & Thoms (2004) and Peat *et al.* (2005) followed the AusRivAS protocol for riffle sampling (Coysh *et al.*, 2000; Peat *et al.*, 2005); whereas Reid *et al.* (2008) sampled benthic macroinvertebrates in different surface flow types, which included habitat types not associated with riffles (Newson & Newson, 2000; Reid & Thoms, 2008). Accordingly, the analyses in this study only utilise assemblage data from Reid & Thoms (2008) collected from surface flow types associated with riffles (chute flow, standing broken waves, standing unbroken waves and ripple flow). In addition, benthic macroinvertebrate assemblages are compared through relative abundances of families and functional feeding groups rather than absolute abundances. Finally, in all cases, only informal analyses are applied to the data and interpretations are made with due consideration of the underlying limitations of making comparisons across the individual studies.

RESULTS

Benthic metabolism varied considerably in space and time (Fig. 2). Overall, P/R ratios were lower before the wildfire and in the first few months after the fire. In both the unregulated and regulated stream there was a general increase in P/R ratios in the months after the fire. Although this change was inconsistent between sites, most site P/R averages from the spring of 2003 onwards (~10 months after the fire) were in excess of 1, and half were more than 2.5 (Fig. 2). Comparisons of trajectories of change in benthic metabolism between unregulated and regulated streams over this period are limited by the lack of data from the unregulated Goodradigbee after spring 2003.

Overall, the number of benthic macroinvertebrate families recorded was higher before the fire than in the samples taken in the first year after the fire for both rivers (Fig. 3). Samples taken in the second year indicated recovery in the number of benthic macroinvertebrate families recorded in the unregulated stream to near pre-fire levels. In contrast, the number of benthic macroinvertebrate families recorded remained low in the unregulated stream in the second year (Fig. 3). After three years, however, the number of benthic macroinvertebrate families recorded in the unregulated stream exceeded those of pre-fire levels in both the regulated and unregulated stream (Fig. 3).

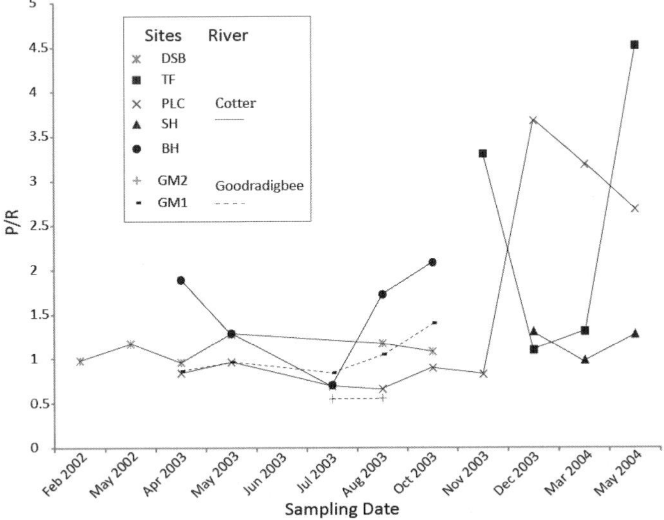

Fig. 2 Primary productivity/respiration ratios (P/R) recorded by Chester (2003), Norris & Thoms (2004) and Reid *et al.* (2006) at various locations on the Cotter and Goodradigbee rivers between February 2002 and May 2004.

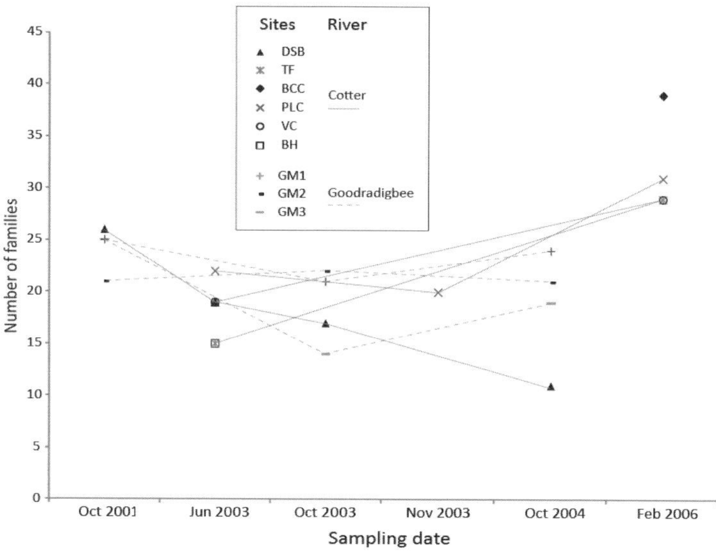

Fig. 3 Number of macroinvertebrate families recorded in riffle samples by Norris & Thoms (2004), Peat *et al.* (2005) and Reid & Thoms (2008) at various locations on the Cotter and Goodradigbee rivers between October 2001 and February 2006.

Assemblage level comparisons could only be made for post-fire macroinvertebrate assemblages in the regulated stream due to the limited amount of raw data available in Peat *et al.* (2005). These comparisons were made using MDS ordinations of family level and functional feeding-group level assemblages (Fig. 4). Both ordinations show clear separation of benthic macroinvertebrates assemblages in samples taken during the first year after the fire from those taken three years after the fire. In the case of the family-level ordination, this separation reflects the higher numbers of oligochaete worms and Caenidae (Ephemeroptera) in assemblages from June and November 2003 and the higher numbers of Gomphidae, Telephlebiidae (Odonata), Leptophlebiidae (Ephmeroptera), Hydropsychidae (Tricoptera) and Corydalidae (Megaloptera) in assemblages sampled in February 2006 (Fig. 4(a)). For the functional feeding group ordination, the separation reflects the higher abundances of "gatherers" and to a lesser extent "shredders" in the 2003 samples, and higher numbers of "scrapers" and "filterers" in the 2006 samples (Fig. 4(b)).

DISCUSSION

It is often difficult to make reliable direct comparisons of data collected by different studies. Taken individually, the benthic metabolism, benthic macroinvertebrate family number and benthic macroinvertebrate assemblage data do not offer a strong basis for making robust inferences about the impact of fire on stream benthic communities in a regulated stream; this is because there is confounding of the factors of interest (pre- *vs* post-fire, regulated *vs* unregulated stream) with the different studies, which each utilised different methods. However, collectively, the data are confirmatory and suggest that the benthic communities that established after 3 years following the fire in the regulated stream are distinct from those prior to the fire and to those that might have been expected in an unregulated stream. The benthic metabolism results from 2004 show a strongly autotrophic benthic community (Reid *et al.*, 2006); this finding is consistent with the later findings of Reid & Thoms (2008) which showed that the macroinvertebrate communities in this reach of the Cotter River are dominated by scraper taxa (Fig. 5) that feed on periphyton communities. Thus, multiple lines of evidence suggest the system is supported by high levels of periphyton production. Increasingly, multiple lines of evidence are being employed in river

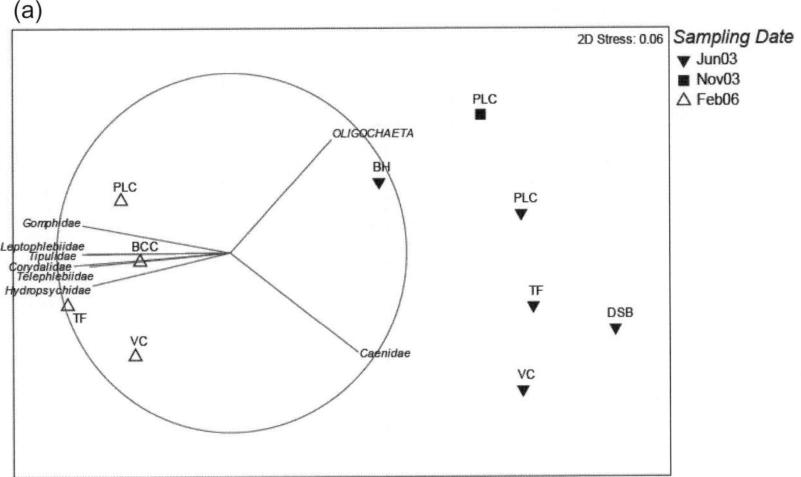

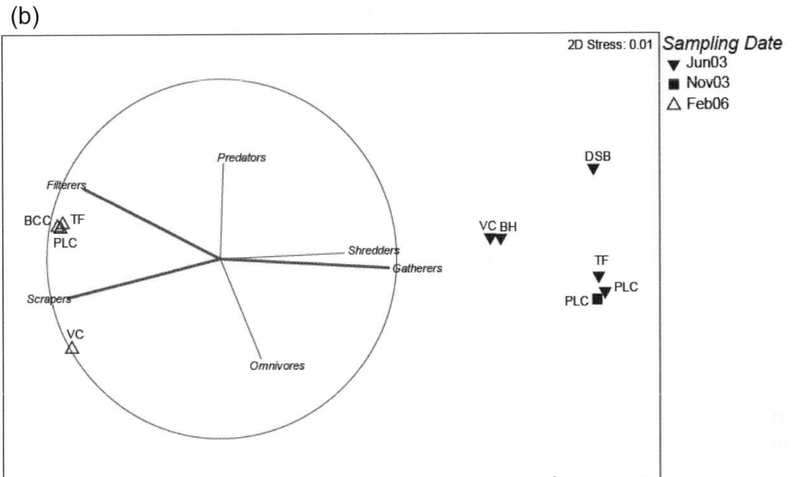

Fig. 4 Multi-Dimensional Scaling (MDS) ordinations of benthic macroinvertebrate assemblages recorded in riffle samples by Norris & Thoms (2004), and Reid & Thoms (2006) at various locations on the Cotter River between June 2003 and February 2006. Ordinations are based on resemblance matrices of Bray-Curtis similarity indices calculated using relative abundances of families (a) and functional feeding groups (b). Vectors indicate the direction of Pearson correlations >0.8 for individual taxa (a) and >0.5 (thin lines) and >0.8 (thick lines) for individual functional feeding groups (b).

science, particularly when exploring ecosystem responses to complex arrays of direct and indirect drivers (Norris *et al.*, 2004) associated with pressures on catchments and water resources.

The Cotter benthic periphyton community appears to have taken some time to become established following the fire. The early post-fire assemblages recorded in the Cotter River in 2003 were similar to those that have been recorded in other streams subject to fire where high catchment runoff and sediment inputs resulted in an abundance of taxa such as chironomids and oligochaete worms that are associated with high concentrations of fine sediment (Mihuc & Minshall, 1995; Minshall, 2003, Norris & Thoms, 2004, Vieira *et al.*, 2004, Peat *et al.*, 2005). However, since this period, it would seem that much of this fine material has been removed from riffle habitats leaving an armoured, highly stable substrate (Southwell *et al.*, 2012; Thoms, 2012). Coupled with reduced

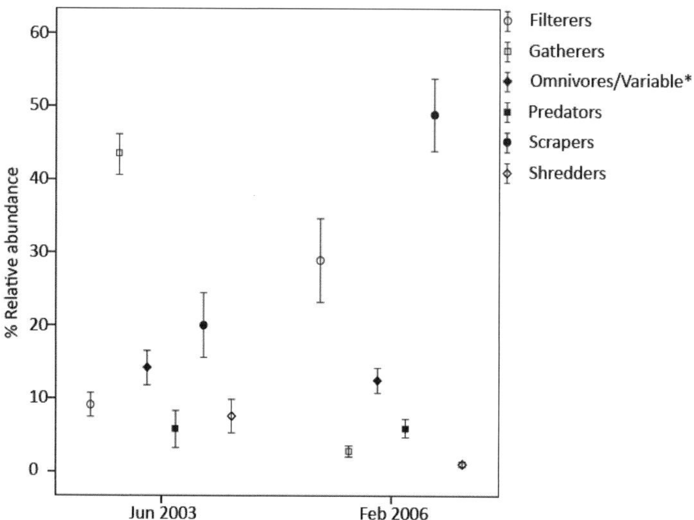

*Families that include taxa with a range of feeding ecologies.

Fig. 5 Relative abundances of macroinvertebrate functional feeding groups recorded in riffle samples by Norris & Thoms (2004), and Reid & Thoms (2006) at various locations on the Cotter River in June 2003 and February 2006. Error bars are standard errors of the means.

shading caused by the opening of the riparian forest canopy and, potentially, high influx of dissolved nutrients from the disturbed catchment (Minshall, 2003; Townsend & Douglas, 2004), these conditions would appear to have facilitated a productive periphyton community that, in turn, supports a large biomass of scraper and filterer macroinvertebrates.

The apparently strongly autotrophic nature of the benthic communities in the Cotter River three years after the fire can be partly attributed to high primary productivity, but may also reflect relatively low levels of respiration in the disturbed system. The quantity and quality of allochthonous material is likely to have been greatly affected by the fire. While post fire influxes of sediment and organic allochthonous materials are generally higher due to increased runoff and availability of material for mobilisation and delivery to river channels, much of the organic debris is likely to be refractory charcoal (Minshall *et al.*, 2001). Moreover, allochthonous organic material may not be retained for long periods due to the armouring of the stream bed (Petts, 1988). The armouring of the bed may also reduce interaction with hyporheic communities where much of the respiration occurs (Petts, 1988).

The results of this study indicate that the interaction of the indirect effects of fire with the effects of flow regulation have resulted in an altered recovery trajectory following the disturbance created by the extensive wildfire in the Cotter River catchment. While it is possible that forest succession and the associated gradual increase in shading of the stream may result in reduced benthic primary productivity in the system, the effect of bed armouring in reducing organic matter retention and fluxes to, and from, the hyporheic zone are likely to persist and thus contribute to the maintenance of an autotrophic system.

CONCLUSION

This study, synthesising the results of several previous projects that examined benthic communities in a regulated stream impacted by wildfire, has shown that the effects of wildfire are likely to be modified under regulated flow regimes. Under natural flow regimes, the principal effects of wildfire relate to the subsequent hydrological disturbances, which deliver high peak flows, high volumes of sediment and changes to channel morphology. Under the regulated flow regime of the Cotter River, these disturbances appear to have been mitigated to a large degree and, as a result, the principal effect of wildfire was a shift to a strongly autotrophic benthic community supported by reduced shading and a stable, armoured substrate that limited interaction with the hyporheos. However, we stress that, given the possible confounding associated with the use of data from several different studies, these conclusions require testing in integrated Before–After/Control–Impact (BACI) studies.

Acknowledgements This manuscript is dedicated to the memory of Prof. Richard Norris, who, through his enthusiasm and dedication to high quality research, made an enormous contribution to stream ecology in Australia.

REFERENCES

Beschta, R. L., Rhodes, J. J., Kauffman, J. B., Griesswell, R. E., Minshall, G. W., Karr, J. R., Perry, D. A., Hauer, E. R. & Frissell, C. A. (2004) Postfire management on forested public lands of the western United States. *Conservation Biology* 18(4), 957–967.

Chester, H. L. (2003) Dams and flow in the Cotter River: effects on instream trophic structure and benthic metabolism. Report, School of Resource, Environment and Heritage Sciences. Canberra, University of Canberra.

Coysh, J. L., Nichols, S. J., Simpson, J. C., Norris, R. H., Barmuta, L. A., Chessman, B. C. & Blackman, P. (2000) Australian River Assessment System (AUSRIVAS) National River Health Program Predictive Model Manual. Co-operative Research Centre for Freshwater Ecology, Canberra, Australia.

Fernandez, C., Vega, J. A., Bara, S., Alonso, M. & Fonturbel, T. (2011) Effects of the sequence wildfire–clearcutting–thinning on nutrient export via streamflow in a small E. globulus watershed in Galicia (NW Spain). *Forest Systems* 20(2), 218–227.

Malison, R. L. & Baxter, C. V. (2010) Effects of wildfire of varying severity on benthic stream insect assemblages and emergence. *J. North American Benthological Society* 29(4), 1324–1338.

McRae, R. (2003) The 2003 ACT bushfires – a review. Curtin, Australian Capital Territory, ACT Emergency Services Bureau: 12.

Mellon, C. D., Wipfli, M. S. & Li, J. L. (2008) Effects of forest fire on headwater stream macroinvertebrate communities in eastern Washington, USA. *Freshwater Biology* 53(11), 2331–2343.

Mihuc, T. B. & Minshall, G. W. (1995) Trophic generalists vs trophic specialists – implications for food-web dynamics in postfire streams. *Ecology* 76(8), 2361–2372.

Minshall, G. F., Brock, J. T., Andrews, D. A. & Robinson, C. T. (2001) Water quality, substratum and biotic responses of five central Idaho (USA) streams during the first year following the Mortar Creek fire. *Int. J. Wildland Fire* 10(2), 185–199.

Minshall, G. W. (2003) Responses of stream benthic macroinvertebrates to fire. *Forest Ecology and Management* 178(1-2), 155–161.

Newson, M. D. & Newson, C. L. (2000) Geomorphology, ecology and river channel habitat: mesoscale approaches to basin-scale challenges. *Progr. Phys. Geogr.* 24(2), 195–217.

Norris, R. & Thoms, M. (2004) Ecological sustainability of modified environmental flows in ACT rivers during the drought 2002-2004. Final Report, May 2004. Canberra, Cooperative Research Centre for Freshwater Ecology: 84.

Norris, R., Liston, P., Mugodo, J., Nicols, S., Quinn, G., Cottingham, P., Metzeling, L., Perriss, S., Robinson, D., Tiller, D. & Wilson, G. (2004) *Multiple Lines and Levels of Evidence for Detecting Ecological Responses to Management Intervention.* 4th Australian Stream Management Conference, Launceston, Tasmania, Tasmanian Department of Primary Industries, Water and Environment.

Peat, M., Chester, H. & Norris, R. (2005) River ecosystem response to bushfire disturbance: interaction with flow regulation. *Australian Forestry* 68(3), 153–161.

Petts, G.E. (1988) Accumulation of fine sediment within the substrate gravels along two regulated rivers, UK. *Regulated Rivers: Research and Management* 2, 141–153.

Reid, M. A., Thoms, M. & Dyer, F. (2006) Effects of spatial and temporal variation in hydraulic conditions on metabolism in cobble biofilm communities in an Australian upland stream. *J. North American Benthological Society* 25(4), 756–767.

Reid, M. A. & Thoms, M. C. (2008) Surface flow types, near-bed hydraulics and the distribution of stream macroinvertebrates. *Biogeosciences* 5, 1043–1055.

Rhoades, C. C., Entwistle, D. & Butler, D. (2011) The influence of wildfire extent and severity on streamwater chemistry, sediment and temperature following the Hayman Fire, Colorado. *Int. J. Wildland Fire* 20(3), 430–442.

Robinson, C. T., Uehlinger, U. & Minshall, G. W. (2005) Functional characteristics of wilderness streams twenty years following wildfire. *Western North American Naturalist* 65(1), 1–10.

Silins, U., Stone, M., Emelko, M. B. & Bladon, K. D. (2009) Sediment production following severe wildfire and post-fire salvage logging in the Rocky Mountain headwaters of the Oldman River Basin, Alberta. *Catena* 79(3), 189–197.

Smith, H. G., Sheridan, G. J., Lane, P. N. J., Noske, P. J. & Heijnis, H. (2011a) Changes to sediment sources following wildfire in a forested upland catchment, southeastern Australia. *Hydrol. Processes* 25(18), 2878–2889.

Smith, H. G., Sheridan, G. J., Lane, P. N. J., Nyman, P. & Haydon, S. (2011b) Wildfire effects on water quality in forest catchments: A review with implications for water supply. *J. Hydrol.* 396(1-2), 170–192.

Smith, H. G., Sheridan, G. J., Lane, P. N. J. & Sherwin, C. B. (2010) Paired Eucalyptus forest catchment study of prescribed fire effects on suspended sediment and nutrient exports in south-eastern Australia. *Int. J. Wildland Fire* 19(5), 624–636.

Southwell, M. & Thoms, M. (2012) Double trouble: the influence of wildfire and flow regulation on fine sediment accumulation in the Cotter River, Australia. In: *Wildfire and Water Quality: Processes, Impacts and Challenges* (Proc. conference held in Banff, Canada, June 2012). IAHS Publ. 354, IAHS Press. Wallingford, UK (this volume).

Thoms, M. C. (2012) The issue below the surface: changes in riverbed sediment structure following wildfires. In: *Wildfire and Water Quality: Processes, Impacts and Challenges* (Proc. conference held in Banff, Canada, June 2012). IAHS Publ. 354, IAHS Press. Wallingford, UK (this volume).

Townsend, S. A. & Douglas, M. M. (2004) The effect of a wildfire on stream water quality and catchment water yield in a tropical savanna excluded from fire for 10 years (Kakadu National Park, North Australia). *Water Research* 38(13), 3051–3058.

Vieira, N. K. M., Clements, W. H., Guevara, L. S. & Jacobs, B. F. (2004) Resistance and resilience of stream insect communities to repeated hydrologic disturbances after a wildfire. *Freshwater Biology* 49(10), 1243–1259.

Wildfire impacts on stream sedimentation: re-visiting the Boulder Creek Burn in Little Granite Creek, Wyoming, USA

SANDRA RYAN & KATHLEEN DWIRE
US Forest Service, Rocky Mountain Research Station, 240 West Prospect Rd., Fort Collins, Colorado 80526, USA
sryanburkett@fs.fed.us

Abstract In this study of a burned watershed in northwestern Wyoming, USA, sedimentation impacts following a moderately-sized fire (Boulder Creek burn, 2000) were evaluated against sediment loads estimated for the period prior to burning. Early observations of suspended sediment yield showed substantially elevated loads (5×) the first year post-fire (2001), followed by less elevated loads in 2002 and 2003, signalling a return to baseline values by 3 years post-fire. However, more recent work (8 years post-fire) has shown elevated suspended sediment yields that are more than double those predicted for the pre-burn range of flows. We tentatively attribute this increase to channel destabilization in the burned area due to the introduction of large wood from burned riparian zones and hillslopes. These results provide insight into the longer-term geomorphic impacts of wildfire that are associated with channel and bank instability in a burned riparian environment due, in part, to large wood dynamics.

Key words post-fire sediment yield; channel instability; instream large wood; Little Granite Creek; Wyoming, USA

INTRODUCTION

In August 2000, a 1400-ha (3500-acre) wildfire in the Gros Ventre Wilderness area near Bondurant, Wyoming, USA, burned substantial portions of the Boulder Creek watershed. Roughly 75% of the forested area in this watershed burned, 67% of which burned at moderate to high severity. The Boulder Creek watershed comprises 40% of the area within Little Granite Creek watershed where researchers from the US Geological Survey (USGS) and US Forest Service (USFS) had previously collected data on bedload and suspended sediment loads (Ryan & Emmett, 2002; Ryan & Dixon, 2008) between 1982 and 1997. The destruction of 1400-ha of old-growth forest in a wilderness area presented an opportunity to quantify increases in sediment loads associated with a wildfire relative to pre-burn baseline values. There are recognized limitations to this approach in that it assumes that the pre-disturbance data sufficiently characterize the flow and sediment regime for the watershed. Yet where adequate pre-disturbance data do exist, the magnitude of the hydrologic and sediment changes can be better quantified by placing flow and sedimentation rates into a broader context through comparisons with background rates (Shakesby & Doerr, 2006). In addition to sediment measurement, a series of monitoring reaches were established to quantify morphological impacts and large wood dynamics in the burned area, areas downstream of the burned area, and within an adjacent reference watershed (Ryan & Dwire, in review). Repeat surveys were used to quantify the changes in channel patterns and cross-sectional area for the first three years post-fire and later in 2007–2008, seven to eight years post-fire.

Results on changes in flow and sediment for the first three years post-fire are presented in Ryan *et al.* (2011). Those earlier observations showed that suspended sediment loads were substantially elevated (5×) the first year post-fire (2001), followed by less elevated loads in 2002 and 2003, suggesting a return to baseline values. In this paper, we present additional results derived from flow and sediment data collected in 2008 to: (a) estimate sediment yield eight years post-fire, and (b) determine whether earlier suppositions on recovery patterns were correct. We also present information on in-stream morphologic changes that are occurring (e.g. bank erosion), due, in part, to the influence of large wood dynamics in the burned riparian zone. Specifics on the dynamics of the large wood are addressed in a separate paper (Ryan & Dwire, in review).

Watershed description

Little Granite Creek (LGC), an upland contributor to the Snake River system, drains 54.6 km^2 (21.1 mi^2) of the Gros Ventre range south of Jackson, Wyoming, USA. The watershed is underlain by deformed sedimentary formations (sandstone and claystone formations of marine origin (Love & Christiansen, 1985) which are relatively unstable in areas. Over 100 active and inactive landslides have been mapped in the watershed, comprising about 20% of the total watershed area (WSGS & WRDS, 2001; refined by S. Ryan using field reconnaissance and aerial photographs). Prior to burning, forest cover was dominated by Lodgepole pine, classified as the persistent Lodgepole pine (*Pinus contorta*) community type, with different understory shrubs and graminoids (Steele *et al.*, 1983; Bradley *et al.*, 1992). In unconfined valley bottoms, riparian vegetation is primarily willow species (*Salix* spp.), with an extensive herbaceous understory (Youngblood *et al.*, 1985). In more confined portions, the flood plain overstory is composed of a mixture of Lodgepole pine, Engelmann spruce (*Picea engelmannii*) and subalpine fir (*Abies lasiocarpa*), with riparian shrubs occurring along the stream banks (Youngblood *et al.*, 1985). Runoff is generated primarily by melting of the annual snowpack, with peak flow occurring between mid-May and mid-June. High, out of bank flows often last 1–2 weeks (USGS, 2007). Mean annual temperature is 1.0°C (33.3°F) and mean annual precipitation is 52.15 cm (20.53 in.) at a climate station in the vicinity of Bondurant (elevation 1982 m/6504 ft) (WRCC, 2005b). Most of the precipitation falls as snow from November through to March. Average annual snowfall measured at Bondurant between 1948 and 1999 was 340 cm, or 134 in. (standard deviation ±124 cm, or 48.9 in.) (WRCC, 2005a).

METHODS

Pre-fire suspended sediment samples and flow measurements were made during the course of 10 runoff seasons between 1982 and 1992 at a USGS gauging station at the confluence between Little Granite Creek and Granite Creek (Ryan & Emmett, 2002). Samples were collected by the USGS using depth-integrating samplers (e.g. DH-48; FISP, 2000). Post-fire flows and sediment transport were monitored at this same location in 2001, 2002 and 2003, and those methods and results are described fully in Ryan *et al.* (2011). While circumstances prevented re-establishing the complete suite of monitoring instruments that were used between 2001 and 2003, we were able to monitor turbidity and use this as a surrogate for suspended sediment concentration (SSC). The sensor (DTS-12 Turbidity Sensor™, Forestry Technology Systems) was deployed from 1 June to 15 September and turbidity was measured at 10-min intervals. A CR10x datalogger (Campbell Scientific) was used to log turbidity data. Intermittent grab samples ($n = 23$) were obtained during snowmelt runoff and baseflows using a DH-48 depth-integrating sampler, and these samples were used to calibrate the signal from the sensors and to develop SSC–discharge relationships. Flow stage was measured from 18 May to 15 September using an automated stage recorder (Aquarod™).

Laboratory analyses on the suspended sediment samples were conducted in accordance with standard USGS methods (Guy, 1969). The mass of material carried in suspension was determined by filtering samples through pre-processed filters, drying the content at 35°C for 2 h, then cooling, desiccating, and weighing the filters to determine the change in weight with added sediment. SSC data are expressed in mass per volume of water collected (mg L^{-1}).

DATA ANALYSIS

Runoff

Discharge estimates were obtained using a flow–stage rating curve developed previously for the site (Ryan & Dixon, 2008). Comparisons between readings from an on-site staff plate and data from the stage-recorders were comparable to the previous period, indicating that the flow-stage rating curve had not shifted and was valid for this purpose. Mean daily flows were calculated as an average of the equally timed measurements for each 24-h period. Since the stage recorders were operational only part of the year, an estimation of flow for the ungauged, early season period was needed. A regression model was developed for this purpose using the available partial year record and data from a nearby gauge (USGS 13023000: Greys River above reservoir near Alpine, Wyoming). Flow estimates from the regression analysis were then used to estimate mean daily flow for the critical data gap (18 May–1 June).

Suspended sediment yield

An estimate of annual suspended sediment yield was made using turbidity–SSC–discharge relationships. Because there were no turbidity measurements obtained prior to 1 June, similar measurements collected in 2003 were used to develop an improved regression between turbidity and SSC. The data from 2008 and 2003 largely overlapped, indicating similarity in the relationship and justifying the incorporation of 2003 data into the model. Moreover, the inclusion of the 2003 data improved the stability of the model, particularly for higher discharges. A least-squares, non-linear regression model was developed between SSC and the corresponding turbidity value:

$$SSC_i = 1.64t^{1.05}$$

where SSC_i is the estimate of instantaneous SSC (in mg L^{-1}) and t is the observed turbidity value (in NTU) ($r^2 = 0.98$). For the period prior to installation of the turbidity sensor, a second model was developed using SSC measurements and discharge Q ($r^2 = 0.89$):

$$SSC_i = 1.45e^{0.0259Q}$$

There is greater uncertainty in the estimates of SSC from this model because sediment concentrations are influenced by fluctuations in supply as well as changing flow (Ryan & Dixon, 2008). However, it does provide an estimate of SSC for the unmeasured period (15 May–1 June), and therefore the estimate of annual sediment yield is not unreasonably low due to data gaps that would be calculated as zero values. The SSC estimates were then multiplied by flow estimates obtained from the stage–discharge relationship. Values from each 24-h period were then averaged to obtain mean daily sediment yield (t d^{-1}) for each day of monitoring during snowmelt and base-flow. These daily values were summed to approximate annual suspended sediment yield (t). While only a portion of the annual runoff was monitored, the majority of sediment is transported during this period of high flow and so most of the total load is accounted for in this series of calculations.

Estimates of erosion, sediment volume and sediment yields from channel banks

Areas of substantial bank erosion, lateral channel migration and avulsions in burned reaches were identified from channel surveys using differences in mapped areas between the 2003 and 2007 or 2008 surveys. Given the composite nature of the channel banks (primarily silt and fine sand over a gravel/cobble base), we assume that most eroded bank material would be moved in suspension, rather than as bedload. Sediment volume was estimated by multiplying eroded area by mean bank height, computed as the difference between the long profile near the eroded bank and banktop. The estimated volumes from each erosional area were summed and averaged for the six burned reaches (including two reaches with negligible bank erosion). Assuming a bulk density of 1 to 1.25 g cm^{-3}, the mass of sediment per unit channel length was calculated and extrapolated over the total length of channel in the burned area to predict the total sediment yield from bank erosion and channel avulsion. This value was divided by 4 to account for the fact that the onset of bank erosion may have started as early as 2003. The estimate of annual sediment yield from bank erosion in the burned area was then compared to the estimated annual suspended sediment yield for 2008.

RESULTS

Runoff

Snowpack and discharges measured between 2001 and 2003 were low, with the highest flows having calculated return frequencies of 1.4 to 1.5 years (bankfull or slightly less). These occurred for 2 days during the entire 3 year period (Ryan *et al.*, 2011). By comparison, discharge measured

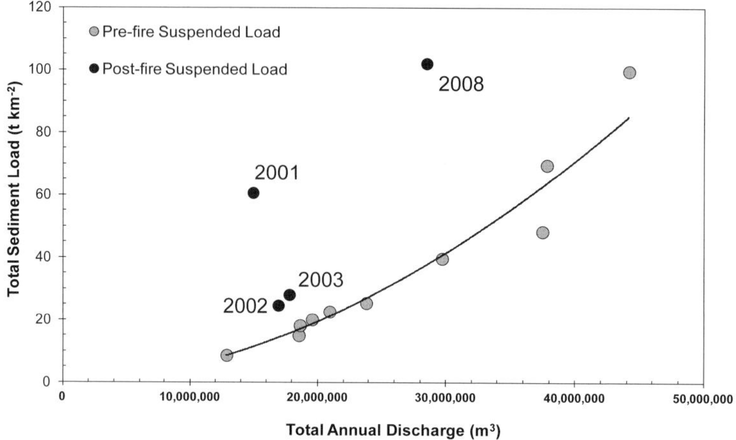

Fig. 1 Annual suspended sediment loads estimated for pre-burn and post-burn periods. Line is fit only to pre-burn data. Values for 2008 may be low as these represent only a partial year estimate (May–September).

in 2008 was relatively high, with out of bank discharges lasting 5 days during the primary peak and a second peak of near bankfull discharges lasting for about a week. Estimated return frequency for peak discharge in 2008 was 2.5 to 3 years. Total daily runoff was summed for the period of observation to approximate annual runoff (Fig. 1). Based on comparisons with the Greys River gauge, we estimate that this accounts for 85% of the total annual flow in 2008 at Little Granite Creek. The estimate of partial annual flow in 2008 was among the highest flows measured at this site.

Estimates of annual suspended sediment load

Suspended sediment loads calculated for 2008 were elevated substantially relative to pre-burn baseline values (Fig. 1). The expected value for the given discharge was 40 t km^{-2} while the calculated value was over 100 t km^{-2}, more than double the pre-burn suspended sediment load. As a caveat, there is some unknown degree of uncertainty in the 2008 estimate, given the limited number of samples collected and restricted period during which the turbidity sensor was deployed. The uncertainty increases for higher flows where the estimate is from the SSC–discharge relationship and no physical samples were obtained. Nevertheless, the models that were used do generate a reasonable approximation of the suspended sediment concentrations for the range of measured calibration samples.

Bank erosion estimates

Between 2003 and 2007, the number of pieces of large wood increased from 303 to 526 in the reaches within the Boulder Creek burn area (mean number of pieces per site increased from 51 to 88). Observed changes in channel form attributed to increased wood input include: (a) channel avulsions; (b) erosion of banks and terraces where wood re-directed flow into the bank; and (c) multiple new sources of fine sediment due to bank instability (Ryan & Dwire, in review). It was postulated that increased channel instability contributed to increased sediment loads measured downstream. Estimates of the volumes of channel bank erosion determined from the mapped areas indicated some between-site variability, but on average, about 12 m^3 of primarily fine sediment was contributed per 100-m sampled reach (Fig. 2). Based on these measurements extrapolated over 12.7 km of stream in the burned area, an estimated 468 t of material may have been available for transport in suspension, annually, from the burned area due to just bank erosion as a sediment

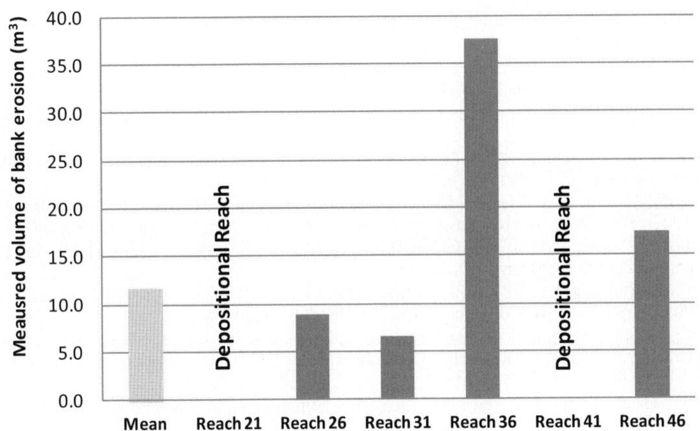

Fig. 2 Measured volume of sediment eroded from channel banks within the burned reaches of Boulder Creek between 2003 and 2007/08.

source. This would account for 15 to 54% of the estimated increase in suspended sediment observed in 2008, depending on whether the erosion occurred evenly over the 4 year timeframe or primarily during the high flow event in 2008.

CONCLUSIONS

This series of observations of increased sediment loads occurring eight years after the Boulder Creek burn in Little Granite Creek watershed suggest that the system is still responding to the large-scale forest disturbance. Calculated annual suspended sediment load was more than double the estimates obtained from the pre-burn period, and proportionally greater than the loads from 2002 and 2003 for the given level of discharge. Although not explicitly addressed in this analysis, we tentatively attribute a portion of these increases to channel destabilization in the burned area, in association with the introduction of large wood from burned riparian areas and hillslopes. This relationship is explained more fully in a separate paper (Ryan & Dwire, in review), but repeat surveys of channel form and large wood distributions show increased bank and bed instability in burned reaches coincident with increased loading of large wood. Estimates of channel bank erosion indicate that up to half of the sediment may be contributed to the annual load from this source. While there is some uncertainty in the estimates of sediment loads and erosion rates, these results provide insight into the longer-term geomorphic impacts of wildfire that are associated with large wood dynamics and changes to channel and bank stability in the burned riparian environment.

REFERENCES

Bradley, A. F., Fischer, W. C. & Noste, N. V. (1992) Fire ecology of the forest habitat types of eastern Idaho and western Wyoming. *USDA For. Serv. INT Res., INT-290.* Ogden, Utah, USA.

FISP (Federal Interagency Sedimentation Project) (2000) Sampling with the US DH-48 depth-integrating suspended-sediment sampler [available on-line (3/22/2008): http://fisp.wes.army.mil/Instructions%20US%20DH-48%20001010.PDF].

Guy, H. P. (1969) Laboratory theory and methods for sediment analysis. *US Geol. Surv. Tech. of Water Res. Invest., Book 5, Chap. C1.*

Love, J. D. & Christiansen, A. C. (1985) *Geologic Map of Wyoming.* USGS, prepared in cooperation with the Wyoming Geologic Survey. 3 sheets.

Ryan, S. E. & Dixon, M. K. (2008) Spatial and temporal variability in stream sediment loads using examples from the Gros Ventre Range, Wyoming, USA. In: *Gravel-Bed Rivers VI: From Process Understanding to River Restoration* (ed. by H. Habersack, H. Piègay, M. Rinaldi), 387–407. Elsevier, Netherlands.

Ryan, S. E. & Dwire, K. A. (in review) Large wood in streams following wildfire and its impact on channel morphology and sediment load. To be submitted to *Geomorphology*.

Ryan, S. E., Dwire, K. A., & Dixon, M. K. (2011) Impacts of wildfire on runoff and sediment loads at Little Granite Creek, western Wyoming. *Geomorphology* 129, 113–130, doi:org/10.1016/j.geomorph.2011.01.017.

Ryan, S. E. & Emmett, W. W. (2002) The nature of flow and sediment movement at Little Granite Creek near Bondurant, Wyoming. *USDA For. Serv. RM Res., RMRS-GTR-90*. Ogden, Utah, USA.

Shakesby, R. A. & Doerr, S. H. (2006) Wildfire as a hydrological and geomorphological agent. *Earth-Sci Rev* 74, 269–307, doi:10.1016/J.EARSCIREV.2005.10.006.

Steele, R., Cooper, S. V., Ondov, D. M., Roberts, D. & Pfister, R. D. (1983) Forest habitat types of eastern Idaho-western Wyoming. *USDA For. Serv. INT Res., INT-144*. Ogden, UT.

USGS National Water Information System: Web Interface USGS 13019438 Little Granite Creek at Mouth nr Bondurant WY. Available on-line (3/22/2007): http://nwis.waterdata.usgs.gov/wy/nwis/nwisman/?site_no=13019438&agency_cd=USGS.

WRCC (Western Regional Climate Center) (2005a) Bondurant, WY Monthly Total Snowfall. Available on-line (7/5/2005): http://www.wrcc.dri.edu/cgi-bin/cliMAIN.pl?wybond.

WRCC (Western Regional Climate Center) (2005b) National Climate Data Center Normals, 1961-90 Bondurant, Wyoming. Available on-line (7/5/2005): http://www.wrcc.dri.edu/cgi-bin/cliNORMNCDC.pl?wybond.

WSGS & WRDS (Wyoming State Geological Survey and the Water Resources Data System) (2001) Preliminary Landslide Map. Wyoming State Geological Survey, Laramie, Wyoming. Available on-line (12/17/2009): http://www.wrds.uwyo.edu/wrds/wsgs/hazards/landslides/lshome.html.

Youngblood, A. P., Padgett, W. G. & Winward, A. H. (1985) Riparian community classification of eastern Idaho-western Wyoming. *USDA For. Serv. R4-Ecol-85-01*.

Application of sediment tracers to discriminate sediment sources following wildfire

HUGH G. SMITH[1], WILLIAM H. BLAKE[1] & PHILIP N. OWENS[2]

1 *School of Geography, Earth and Environmental Sciences, Plymouth University, Plymouth, Devon PL4 8AA, UK*
hugh.smith@plymouth.ac.uk

2 *Environmental Science Program and Quesnel River Research Centre, University of Northern British Columbia, Prince George, British Columbia V2N 4Z9, Canada*

Abstract Wildfire can cause the main sources of fine sediment transported within burned basins to change. Sediment tracing techniques offer an approach to quantify this sediment source response to burning. In this paper, we review the application of various sediment tracers to discriminate sediment sources following wildfire. This includes studies examining changes in fallout radionuclide, geochemical and mineral magnetic properties of burned soil to identify possible fire-related effects on tracer properties and behaviour. Previous work suggests that fallout radionuclides (^{137}Cs, excess ^{210}Pb, ^{7}Be, ^{239,240}Pu) may provide the most effective tracers for post-fire sediment source and budgeting studies. Other tracer properties seem to be overly susceptible to burn effects and might provide a less consistent basis for post-fire source discrimination. The final selection of tracer properties for use in sediment tracing applications should reflect the likely controls of individual tracer concentrations in different sources and their potential response to burning.

Key words wildfire; sediment tracing; sediment sources; fallout radionuclides; river basins

INTRODUCTION

Wildfires represent an important form of disturbance in forest environments that may result in significant changes to runoff and erosion processes, as well as patterns of sediment redistribution in river basins (Shakesby & Doerr, 2006). Fire impacts on erosion and sediment transfer at the basin-scale reflect the balance of sediment inputs from hillslopes and the channel network. Fire-related modifications to soil and surface properties associated with soil heating, loss of vegetative cover and the formation of surface ash deposits can result in substantial increases in hillslope-scale runoff generation and soil erosion (Shakesby & Doerr, 2006). Channel network sediment contributions may also increase in response to elevated post-fire storm flows that are highly erosive (Moody & Martin, 2001). Presently, few studies examine the contributions of fine sediment from major sources to the loadings exported both within, and from, burned river basins. However, there is a need for such information to facilitate better understanding of wildfire impacts on sediment transfer at the basin-scale. Furthermore, this information may also provide the basis for more targeted post-fire mitigation strategies designed to reduce sediment and associated contaminant impacts on water quality. The latter is particularly important for water supply source areas affected by wildfires (Smith *et al.*, 2011a).

Sediment source tracing offers an approach to determine the relative contributions of sediment from potential sources. Source tracing techniques have become well-established over the past two decades (Walling, 2005) and have been widely applied in agricultural catchments. More recently, there has been increased interest in the use of source tracers in forest landscapes following wildfires, particularly in southeastern Australia and western Canada. In this paper, we review and synthesise the applications of sediment source tracers in burned forest environments to inform potential users of the different tracers available. On this basis, the objective of this paper is to examine the application of various tracer properties and their source discrimination potential in burned forest environments. We focus on tracers that may discriminate both sediment source types (e.g. surface and sub-surface or channel bank sources) and spatially-defined sources (e.g. burned *versus* unburned, or areas burned at different severities). Tracer properties include fallout radionuclides (^{137}Cs, excess ^{210}Pb, ^{7}Be, ^{239,240}Pu), soil geochemical properties, mineral magnetics and organic compounds. Fire effects on soils may cause modifications to these properties. Therefore, we consider possible fire-related changes to sediment tracers which may affect their source discrimination potential.

FALLOUT RADIONUCLIDES

Fallout radionuclides have been the most frequently used tracers in post-wildfire sediment source and budgeting studies. The environmental occurrence of those fallout radionuclides used as sediment tracers has been extensively described in previous studies (e.g. Walling & Woodward, 1992; Wallbrink & Murray, 1993; Everett *et al.*, 2008). In brief, ^{137}Cs (half-life 30.2 years) was produced by atmospheric nuclear weapons testing during the 1950s–1960s. Likewise, ^{239}Pu (half-life 24 110 years) and ^{240}Pu (half-life 6561 years) were dispersed globally following nuclear weapons testing. Excess ^{210}Pb ($^{210}Pb_{ex}$) refers to the amount of ^{210}Pb (half-life 22.3 years) derived from natural atmospheric fallout which exceeds that supported directly by decay of its lithogenic parent ^{226}Ra *in situ*. ^{7}Be (half-life 53.3 days) is another naturally occurring radionuclide that is produced by cosmic ray spallation of nitrogen and oxygen in the upper atmosphere. The fallout of each radionuclide is predominantly associated with precipitation. These radionuclides are typically effective sediment source tracers because of their tendency to strongly adsorb to fine particles, pronounced variation in activity concentration with soil depth, and generally conservative behaviour during transport (Wallbrink & Murray, 1993; Wallbrink *et al.*, 1998; Everett *et al.*, 2008). However, some fallout may be lost in surface runoff (Dalgleish & Foster, 1996) which could affect the results of sediment budgeting studies that quantify sediment redistribution based on spatially-defined source zones. Such studies assume fallout is retained in the soil and that radionuclide redistribution occurs only by erosion and deposition. This may lead to erroneous results (Parsons & Foster, 2011) but is less of an issue for source tracing studies because they do not require a reference site to estimate the total fallout inventory. Additionally, preferential radionuclide adsorption to fine particles and selective transport generally produce differences in radionuclide concentrations between source soil and sediment samples that are related to particle size (He & Walling, 1996). This also applies to other tracer properties (Horowitz, 1991). As a result, particle size corrections based on specific surface area ratios or analysis of a very fine fraction (typically <10 μm) have been used to minimise the effect of particle size differences on tracer concentrations (Collins *et al.*, 1997; Wallbrink *et al.*, 1999).

In forested landscapes, maximum concentrations of ^{7}Be and $^{210}Pb_{ex}$ typically occur in surface litter and at the soil surface, respectively, due to the constant production and delivery of these fallout radionuclides. For ^{137}Cs and ^{239,240}Pu, maximum levels tend to be slightly below the surface due to translocation of these fallout radionuclides over the decades since fallout, which is associated with organic matter decay, geochemical diffusion, bioturbation and elluviation processes (Wallbrink & Murray, 1996; Wallbrink *et al.*, 2002; Dong *et al.*, 2010). All of the fallout radionuclides exhibit activity depth profiles that tend to decrease exponentially with depth in undisturbed forest environments. These distinct depth profiles enable fallout radionuclides to readily discriminate surface and sub-surface (e.g. channel bank) sources of sediment based on differences in activity concentrations (e.g. Fig. 1(a)). Several studies have employed ^{137}Cs, $^{210}Pb_{ex}$ and ^{239,240}Pu to examine hillslope and channel bank source contributions to sediment transported both within and exported from burned catchments (Wilkinson *et al.*, 2009; Smith *et al.*, 2011b; 2012; Owens *et al.*, 2012).

Burning has been reported to increase concentrations of ^{137}Cs and $^{210}Pb_{ex}$ in the surface layers of forest soils (Paliouris *et al.*, 1995; Johansen *et al.*, 2003; Wilkinson *et al.*, 2009), and a similar trend may be expected for ^{7}Be and ^{239,240}Pu. This could allow source discrimination of burned and unburned surface soils, depending on the magnitude of increase, its persistence and the relative contribution of ash to the change in radionuclide activity concentrations at the soil surface. The latter relates to the pre-fire distribution of radionuclides between surface vegetation, soil organic matter and mineral soil. For example, Johansen *et al.* (2003) reported that average concentrations of ^{137}Cs were 40 times higher in ash deposits (192 Bq kg^{-1}; range 7–570 Bq kg^{-1}) and three times higher in the upper 50 mm of burned soils (44 ± 2 Bq kg^{-1}) compared to pre-fire soils (15 ± 3 Bq kg^{-1}) under Ponderosa pine near Los Alamos, New Mexico. The elevated ^{137}Cs levels in the ash layer were largely attributed to burning of the pine needle duff layer that had remained unburned for the prior 60 years and which provided a store of ^{137}Cs accumulated from fallout. In contrast to the conifer forest environment, the pattern of fallout radionuclide distribution appears to differ for eucalypt forest soils in southeastern Australia. While increased post-fire concentrations of ^{137}Cs and $^{210}Pb_{ex}$ in surface soils under eucalypt forest have been reported (Wilkinson *et al.*, 2009), ^{137}Cs seems to be less associated with ash in contrast to $^{210}Pb_{ex}$ and ^{7}Be. This reflects the partitioning of $^{210}Pb_{ex}$ and ^{7}Be between surface litter and soil, with up to 31% and 95% of total inventories stored in surface litter in eucalypt forests, respectively, compared to 15% for ^{137}Cs (Wallbrink & Murray, 1996; Wallbrink *et al.*, 1997). ^{137}Cs in eucalypt forest soils is predominantly associated with mineral soil due to bioturbation transferring decayed litter into the soil following the cessation of ^{137}Cs fallout (Wallbrink *et al.*, 2005; Blake *et al.*, 2009). As a consequence, in eucalypt forests, $^{210}Pb_{ex}$ and ^{7}Be tracers may be more representative of ash and surface litter, while ^{137}Cs may better represent mineral and organic soil immediately below the ash layer and any remaining O horizon. The apparent difference in ^{137}Cs levels associated with ash formed from the duff layer in a North American conifer forest compared to surface litter in eucalypt forest may reflect greater ^{137}Cs fallout in the Northern Hemisphere (Hodge *et al.*, 1996) combined with slower rates of litter decay and bioturbation in conifer forests under cooler climatic conditions.

Burning may enhance spatial variability in fallout radionuclide concentrations in surface soils. This reflects spatial variations in fuel load and burn patterns that result in varying levels of ash formation and combustion of soil organic material that affects surface soil radionuclide concentrations. Local redistribution or concentration of ash by wind may also be a factor. There is evidence that $^{210}Pb_{ex}$ concentrations in burned surface soils exhibit greater spatial variability than ^{137}Cs. Smith *et al.* (2012) reported the coefficient of variation (CV) for $^{210}Pb_{ex}$ and ^{137}Cs measurements of combined surface soil and ash samples from two burned eucalypt forest catchments as well CVs calculated from data presented by two other post-fire tracer studies undertaken in eucalypt forests (Wilkinson *et al.*, 2009; Smith *et al.*, 2011b). The CV range for $^{210}Pb_{ex}$ concentrations (0.04–0.39) in burned surface soils was found to generally exceed ^{137}Cs (0.06–0.15). This was attributed to the higher concentration of $^{210}Pb_{ex}$ in eucalypt surface vegetation and litter (Wallbrink *et al.*, 1997) that was transferred to ash, thereby increasing variability in surface soil $^{210}Pb_{ex}$ concentrations relative to ^{137}Cs. Owens *et al.* (2012) examined temporal variability in post-fire radionuclide concentrations by sampling surface soils in the first and fourth years after wildfire in conifer forest in British Columbia. ^{137}Cs and $^{210}Pb_{ex}$ concentrations were not significantly different between the two sampling occasions, although mean concentrations were higher in the first year post-fire. Notably, radionuclide concentrations remained significantly greater ($p < 0.05$) in burned surface soil collected four years after the fire compared to soil collected from a nearby unburned catchment. This suggests that fire may have a persistent effect on radionuclide concentrations in surface soils in some landscapes, with clear implications for source discrimination between burned and unburned areas. Importantly, much of the ^{7}Be in surface material could be delivered after burning, given that fires are generally associated with preceding periods of low rainfall during which the inventory of this short-lived radionuclide would decline to low levels. Therefore, ^{7}Be may be less useful for discriminating burned *versus* unburned areas given that, unlike ^{137}Cs and $^{210}Pb_{ex}$, most of the inventory may be delivered with rainfall after the fire and would not be subject to burn effects.

The effect of fire on fallout radionuclide concentrations and their distribution between ash and surface soils may assist post-fire sediment tracing applications. For source tracing investigations that seek to apportion contributions from hillslope surface and sub-surface (usually channel bank) sources, the tendency for radionuclide concentration to increase in burned surface soils may improve source discrimination. However, the potential for increased spatial variability in radionuclide concentrations from burned surface soils may also contribute to greater uncertainty in estimates of surface source contributions to post-fire sediment flux. Burning does not appear to affect sub-soil radionuclide concentrations based on comparison of measurements of sub-surface soils (>5 cm) and eroding channel banks from burned and unburned catchments (Owens *et al.*, 2012). The partitioning of radionuclides between ash/surface litter ($^{210}Pb_{ex}$, ^{7}Be) and mineral soil (^{137}Cs) observed in eucalypt forests has been used to develop post-fire sediment budgets, based on each radionuclide, that capture the contrasting erosion response of the ash/surface organic layer *versus*

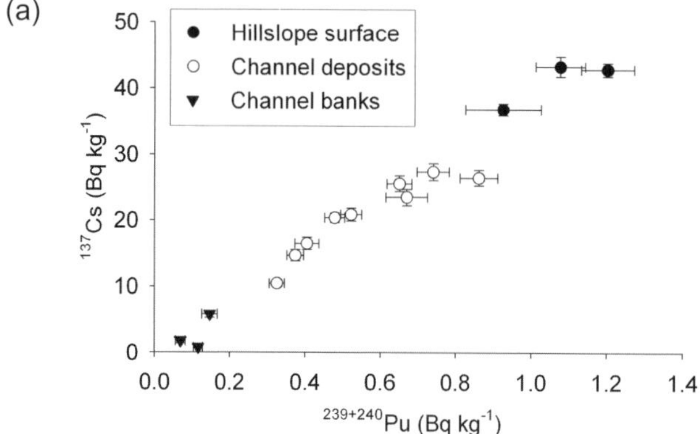

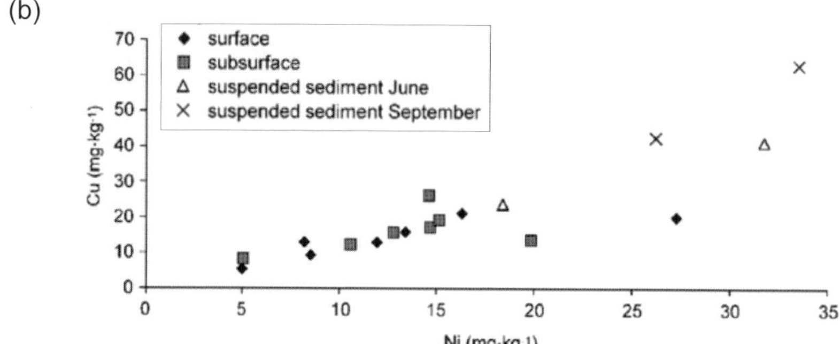

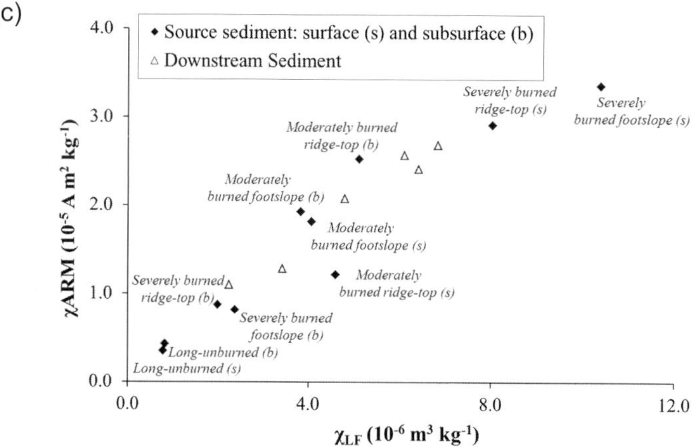

Fig. 1 Examples of bivariate plots comparing various tracer properties from source material with deposited or suspended sediment for: (a) fallout radionuclides ^{137}Cs and $^{239+240}$Pu in composite samples (<10 μm fraction) collected from small catchments affected by post-fire debris flows in northeastern Victoria, Australia (after Smith *et al.*, 2012); (b) geochemical constituents (Cu and Ni) in the <500 μm fraction (with particle size correction between source and sediment samples) from a burned conifer catchment in British Columbia, Canada, in the first year after fire (after Owens *et al.*, 2006); and (c) mineral-magnetic properties associated with soil under eucalypt forest burned at different severities as well as river sediment deposits (all <10 μm fraction) in a catchment near Sydney, Australia (after Blake *et al.*, 2006b).

the underlying mineral soil (Wallbrink *et al.*, 2005; Blake *et al.*, 2009). The different half-lives and fallout patterns of these radionuclides also provide a basis for examining sediment budgets over timescales ranging from months (immediate post-fire sediment redistribution based on ^{7}Be and ^{210}Pb$_{ex}$) to decades (longer-term patterns of sediment transfer using ^{137}Cs) (Blake *et al.*, 2009).

The presence of fallout radionuclides in ash may have implications for estimating surface source contributions to post-fire catchment sediment exports. The deposited ash layer is transient and may be largely removed within the first year after fire, or sooner, depending on the erosivity of post-fire rainfall and wind events (Reneau *et al.*, 2007). Some proportion of ash may also be progressively incorporated into the surface soil through infiltration and mixing associated with bioturbation. As a result, the relative contribution of ash to post-fire catchment exports will change with time since fire. This may present a challenge when characterising surface source radionuclide concentrations based on a single sampling occasion after fire. Changes in the composition (i.e. ash content) of surface material exported from the catchment over time could affect radionuclide concentrations relative to surface material *in situ* comprised of a mixture of ash and soil. The potential significance of this temporal change in surface source signatures will depend on the magnitude of the change relative to spatial variability in tracer properties from the initial sample collection. Given that ^{137}Cs and ^{210}Pb$_{ex}$ concentrations in surface soils have been observed not to be significantly different between the first and fourth years after wildfire (Owens *et al.*, 2012), temporal changes might be of less importance than initial post-fire spatial variability. However, repeated surface source sampling could still be required in some instances and will be essential if ^{7}Be is used in post-fire tracing applications due to its short half-life and ongoing fallout.

GEOCHEMICAL PROPERTIES

Soil geochemical properties have frequently been incorporated in multi-tracer fingerprinting studies of sediment sources, mainly in agricultural or mixed agricultural and urban catchments (e.g. Walling, 2005). The range of potential tracer properties used typically includes various metals (such as Al, Cd, Co, Cr, Cu, Fe, Mn, Ni, Pb, Sr, Zn), base cations (Ca, K, Mg, Na) and nutrients (N, P, C) (Walling *et al.*, 2008). The selection of a specific geochemical property as part of the final set of tracers used in source ascription is dependent upon establishing statistical differences between target sources for a given study catchment (Collins & Walling, 2002). The potential causes of these differences may relate to natural or human-impacted processes and controlling factors affecting the development of soil profiles. For example, differences in soil geochemical signatures may be influenced by differences in soil parent material, the degree of weathering or by modifications to the soil surface such as by cultivation or burning. Soil nutrients may act as tracers of surface and sub-surface soil based on higher concentrations in surface soils from vegetation decomposition and nutrient cycling.

Soil heating by wildfire can modify surface soil geochemistry (Chambers & Attiwill, 1994; Certini, 2005), thereby resulting in the potential for source discrimination of severely burned, moderately burned and unburned soils (Blake *et al.*, 2006a). For example, levels of total Mn in soils have been reported to increase following fire through additions in ash and the physiochemical breakdown of Mn complexed with organic matter (Chambers & Attiwill, 1994), and similar changes might be expected for Fe, Cu, and Zn (Certini, 2005). In contrast, burning has been observed to reduce surface soil total N (Chambers & Attiwill, 1994; Murphy *et al.*, 2006), while total P generally does not decrease after wildfire because the volatilization temperature of P (>550°C) is much greater than N (200°C) (Raison, 1979; Hernandez *et al.*, 1997; Murphy *et al.*, 2006).

The use of soil geochemical properties to distinguish sediment sources in post-fire environments has been explored in eucalypt forest in the Blue Mountains near Sydney, Australia (Blake *et al.*, 2006a). In this study, surface soil samples collected from ridge-tops characterized by severe and moderate burn severities as well as unburned conditions were measured for Ca, Mg, P, Pb and Zn using X-ray fluorescence. Blake *et al.* (2006a) found that there was an increase in concentrations of elements in burned relative to unburned soil, with the highest levels reported in the severely burned soil. The increase in concentrations with burn severity was attributed to progressive mineralization of organic matter, such that the organic mass declined in soils burned at higher severities and resulted in increased concentrations of the various constituents measured. These authors concluded that burn effects on soil geochemistry might allow source discrimination of areas of differing burn severity but the "tagged" material largely comprised ash.

Following wildfire in a conifer forest catchment in British Columbia, Owens *et al.* (2006) reported that fire-induced changes to various soil geochemical properties both enhanced and compromised their potential as sediment source tracers. This study focused on the discrimination potential of geochemical tracers between surface (topsoil) and sub-surface (including channel banks) sources (Fig. 1(b)). Various geochemical properties were significantly different between the two sources in both the burned and unburned study catchments. In comparison with the unburned samples, burning resulted in both increases (Co, Cu, K, Ni, P, Zn) and decreases (C, Mn, Pb) in constituents measured in surface material. However, bivariate plots of C *versus* N indicated that a reduction in C concentrations in surface soils, combined with elevated concentrations (exceeding surface soils) in post-fire suspended sediment samples, confounded its use as a source tracer (Owens *et al.*, 2006). The increased C concentrations in suspended sediment may reflect in-stream modifications, such as biofilm development (Petticrew *et al.*, 2006). Similarly, comparison of Cu and Ni between sources and suspended sediment in both burned and unburned catchments (Fig. 1(b)) did not provide the basis for effective discrimination of surface and sub-surface source contributions to suspended sediment. These findings emphasise the need to consider both background and fire-related controls of tracer properties as well as potential in-stream modifications.

MINERAL MAGNETICS

Mineral magnetic properties are often included as part of multi-parameter tracing investigations (e.g. Russell *et al.*, 2001). The most common types of magnetic minerals are haematite, goethite, maghemite and magnetite (Smith, 1999) and, in soil, they originate from both natural and anthropogenic processes (Oldfield, 1991, 1999). Magnetic minerals can occur as discrete fine grains, aggregated concretions or very fine-grained particle coatings (Taylor *et al.*, 1983). From a source tracing perspective, the origin of the magnetic signal is important. Minerals derived from parent material contribute to the "primary" magnetic signature of the soil, which is often overprinted by "secondary" minerals of pedogenic, bacterial, anthropogenic or pyrogenic origin. The sensitivity of secondary magnetic mineral assemblages to short and longer-term environmental change offers potential for sediment source ascription in burned landscapes. The use of mineral magnetic properties as sediment tracers in burned landscapes is based on the production of fine-grained ferrimagnetic minerals in surface soil material during burning (Rummery *et al.*, 1979). Pyrogenic enhancement of soil magnetic properties is linked to changes in oxidation/reduction conditions in surface soil during burning which convert less magnetic iron oxy-hydroxides into more magnetic minerals (Blake *et al.*, 2006c).

The presence of pyrogenic ferrimagnetic minerals, such as magnetite or maghemite, is reflected in enhanced magnetic susceptibility (χ) and saturation isothermal remanent magnetization (SIRM) measurements which are indicative of increased magnetic concentration in burned soil (Oldfield & Crowther, 2007). Mineral magnetic signatures of burned material can be further developed and quantified by looking at magnetic grain-size dependent properties, for example frequency dependent magnetic susceptibility (χ_{fd}) and susceptibility of anhysteritic remanent magnetisation (χ_{ARM}) (Blake *et al.*, 2006b). Quotients of these measures i.e. χ_{ARM}/χ_{lf} or χ_{ARM}/χ_{fd} can be used to distinguish source materials and downstream sediment, and can be usefully compared in bivariate diagrams wherein distinctive envelopes of values due to burning can be identified (Oldfield & Crowther, 2007). Quotients, however, are generally considered to be non-linearly additive (Lees, 1999) which precludes them from use in quantitative source unmixing models.

Recent studies (Blake *et al.* 2006c) have demonstrated spatial variability in mineral magnetic signatures linked to fire severity (Fig. 1(c)). The similarity between the temperature thresholds of magnetic enhancement and changes in soil hydrological properties (e.g. soil water repellency, Doerr *et al.*, 2004) offers potential to link sediment sources related to burn severity to fire impacts on hillslope hydrology (Blake *et al.*, 2006c). Use of concentration-based parameters in this regards shows promise but in many cases source end members represent numerical multiples of each other (Fig. 1(c)), emphasising the need to involve additional properties and/or complementary monitoring or field survey data to allow a more robust discrimination (Blake *et al.*, 2006b).

ORGANIC COMPOUNDS

The use of organic compounds as sediment source tracers is receiving increasing attention with developments in analytical technology. Both the molecular composition of organic compounds and the stable isotope (C and N) signatures of organic matter offer potential in this regard. Sediment tracing applications are in their infancy but recent studies have demonstrated the potential of (i) soil enzyme activities to trace stream sediment to sources under different land use (e.g. Nostrati *et al.*, 2011), (ii) bulk C and N isotopic ratios to explore catchment sediment sources (e.g. Fox & Papanicolaou, 2008), and (iii) Compound Specific Stable Isotope (CSSI) signatures of sediment to explore estuarine sediment sources liked to forestry operations (e.g. Gibbs, 2008) and to link stream sediment to fields of specific crop type (e.g. Blake *et al.*, 2012). Burning of biomass offers potential for further modification of organic compound composition in soil (Atanassova & Doerr, 2011) but the potential for sediment tracing remains largely unexplored. A notable exception in the context of burned soils is a study by Oros *et al.* (2002) which examined the molecular composition of plant and soil organic compounds before and after controlled and wildfire burning. They noted that heating and post-fire leaching and bio-transformations were all important factors in developing organic geochemical fingerprints of burned soils. They concluded that organic geochemical analysis can yield useful chemical fingerprints that identify plant material contributing to organic matter in burned soils and, by extension, this could be a useful soil and sediment tracer. Despite the promise shown by this study, there have been no further studies in this area, perhaps due to the complexity and associated costs of the analyses.

PERSPECTIVE

The application of tracers to investigate sources of fine sediment within burned catchments has considerable potential to contribute to the understanding of post-fire erosion processes and sediment transfer across a range of basin scales and forest environments. However, there are several issues relating to the use of sediment tracers in burned environments that warrant further consideration. These include, amongst others, the need for improved spatial characterisation of tracer properties in source materials. Higher levels of spatial variability in tracer properties from burned forest soils than in other landscape types may necessitate collection of greater numbers of source material samples as well as the use of a composite sampling strategy. In addition, more information is required on tracer partitioning between ash, litter/duff and mineral soil. There is also a need to determine the extent to which tracer properties in burned soil change over time and related to this, repeated post-fire source sampling may be necessary. Lastly, some tracer properties may undergo geochemical alteration as they move between landscape compartments, especially from terrestrial to aquatic systems. Further work investigating the effect of in-stream processes on tracer behaviour and properties is required.

Acknowledgements This review was written with the support of a Marie Curie International Incoming Fellowship (HGS) within the 7th European Community Framework Programme. WB acknowledges UK NERC (NER/A/S/2002/00143; NE/F01273X/1) and Royal Society funding of post-fire hydrological studies. PNO received financial support via an NSERC Discovery Grant and a Forest Renewal BC Operating Grant for work in British Columbia, and would like to thank Tim Giles (BC Ministry of Forests and Range) for support with fieldwork over the last 8 years. The draft manuscript was improved by review comments.

REFERENCES

Blake, W. H., Wallbrink, P. J., Doerr, S. H., Shakesby, R. A., Humphreys, G. S., English, P. & Wilkinson, S. (2006a) Using geochemical stratigraphy to indicate post-fire sediment and nutrient fluxes into a water supply reservoir, Sydney, Australia. In: *Sediment Dynamics and the Hydromorphology of Fluvial Systems* (ed. by J. S. Rowan, R. W. Duck & A. Werritty), 363–370. IAHS Publ. 306, IAHS Press, Wallingford, UK.

Blake, W. H., Wallbrink, P. J., Doerr, S. H., Shakesby, R. A. & Humphreys, G. S. (2006b) Magnetic enhancement in wildfire-affected soil and its potential for sediment-source ascription. *Earth Surface Processes and Landforms* 31, 249–264.

Blake, W. H., Doerr, S. H., Wallbrink, P. J., Shakesby, R. A., Humphreys, G. S. & Chafer, C. (2006c). Tracing eroded soil in a burned water supply catchment, Sydney, Australia: linking magnetic enhancement to soil water repellency. In: *Soil Erosion and Sediment Redistribution in River Catchments: Measurement, Modelling and Management* (ed. by P.N. Owens & A.J. Collins), 62–69. CAB International.

Blake, W. H., Wallbrink, P. J., Wilkinson, S. N., Humphreys, G. S., Doerr, S. H., Shakesby, R. A. & Tomkins, K. M. (2009) Deriving hillslope sediment budgets in wildfire-affected forests using fallout radionuclide tracers. *Geomorphology* 104, 105–116.

Blake, W. H., Ficken, K. J., Taylor, P., Russell, M. A. & Walling, D. E. (2012) Tracing crop-specific sediment sources in agricultural catchments,.*Geomorphology* 139-140, 322–329

Caitcheon, G. G. (1993) Sediment source tracing using environmental magetism: a new approach with examples from Australia. *Hydrol. Processes* 7, 349–358.

Certini, G. (2005) Effects of fire on properties of forest soils: a review. *Oecologia* 143, 1–10.

Chambers, D. P. & Attiwill, P. M. (1994) The ash-bed effect in *Eucalyptus regnans* forest: chemical, physical and microbiological changes in soil after heating or partial sterilisation. *Australian Journal of Botany* 42, 739–749.

Collins, A. L., Walling, D. E. & Leeks, G. J. L. (1997). Source type ascription for fluvial suspended sediment based on a quantitative composite fingerprinting technique. *Catena* 29, 1–27.

Collins, A. L. & Walling, D. E. (2002) Selecting fingerprint properties for discriminating potential suspended sediment sources in river basins. *J. Hydrol.* 261, 218–244.

Dalgleish, H. Y. & Foster, I. D. L. (1996) ^{137}Cs losses from a loamy surface water gleyed soil (Inceptisol); a laboratory simulation experiment. *Catena* 26, 227–245.

Doerr, S. H., Blake, W. H., Shakesby, R. A., Stagnitti, F., Vuurens, S. H., Humphreys, G. S. & Wallbrink, P. (2004) Heating effects on water repellency in Australian eucalypt forest soils and their value in estimating wildfire soil temperatures. *Int. J. Wildland Fire* 13, 157–163.

Dong, W., Tims, S. G., Fifield, L. K. & Guo, Q. (2010) Concentration and characterization of plutonium in soils of Hubei in central China. *J. Environmental Radioactivity* 101, 29–32.

Everett, S. E., Tims, S. G., Hancock, G. J., Bartley, R. & Fifield, L. K. (2008) Comparison of Pu and ^{137}Cs as tracers of soil and sediment transport in a terrestrial environment. *J. Environmental Radioactivity* 99, 383–393.

Fox, J. F. & Papanicolaou, A. N. (2008) Application of the spatial distribution of nitrogen stable isotopes for sediment tracing at the watershed scale. *J. Hydrol.* 358, 46–55.

He, Q. & Walling D.E. (1996). Interpreting particle size effects in the adsorption of ^{137}Cs and unsupported ^{210}Pb by mineral soils and sediments. *J. Environmental Radioactivity* 30, 117–137.

Hernandez, T., Garcia, C. & Reinhardt, I. (1997) Short-term effect of wildfire on the chemical, biochemical, and microbiological properties of Mediterranean pine forest soils. *Biology and Fertility of Soils* 25, 109–116.

Hodge, V., Smith C. & Whiting J. (1996) Radiocesium and plutonium: still together in "background" soils after more than thirty years. *Chemosphere* 32, 2067–2075.

Horowitz, A. J. (1991) *A Primer in Sediment-trace Element Chemistry*, 2nd edn. Lewis Publishers: Chelsea, Michigan, USA.

Johansen, M.P., Hakonson, T.E., Whicker, F.W. & Breshears, D.D. (2003) Pulsed redistribution of a contaminant following forest fire: cesium-137 in runoff. *J. Environmental Quality* 32, 2150–2157.

Lees, J. (1999) Evaluating magnetic parameters for use in source identification, classification and modelling of natural and environmental materials. In: *Environmental Magnetism: A Practical Guide* (ed. by J. Walden, F. Oldfield & J. Smith), 113–138. Quaternary Research Association.

Moody, J. A. & Martin, D. A. (2001) Initial hydrologic and geomorphic response following a wildfire in the Colorado Front Range. *Earth Surface Processes and Landforms* 26, 1049–1070.

Murphy, J. D., Johnson, D. W., Miller, W. W., Walker, R. F., Carroll, E. F. & Blank, R. R. (2006) Wildfire effects on soil nutrients and leaching in a Tahoe Basin watershed. *J. Environmental Quality* 35, 479–489.

Nosrati, K., Govers, G., Ahmadi, H., Sharifi, F., Amoozegar, M. A., Merckx, R. & Vanmaercke, M. (2011) An exploratory study on the use of enzyme activities as sediment tracers: biochemical fingerprints? *Int. J. Sediment Research* 26, 136–151.

Oldfield, F. (1991) Environmental magnetism: a personal perspective. *Quaternary Science Reviews* 10, 73–85.

Oldfield, F. (1999) The rock magnetic identification of magnetic mineral and grain size assemblages. In *Environmental Magnetism: A Practical Guide, (*ed. by J. Walden, F. Oldfield & J. Smith), 98–112. Quaternary Research Association.

Oldfield, F. & Crowther, J. (2007) Establishing fire incidence in temperate soils using magnetic measurements. *Palaeogeography, Palaeoclimatology, Palaeoecology* 249, 362–369.

Oros, D. R., Mazurek, M. A., Baham, J. E. & Simoneit, B. R. T. (2002) Organic tracers from wildfire residues in soils and rain/river wash-out. *Water, Air, and Soil Pollution* 137, 203–233.

Owens, P. N., Blake, W. H. & Petticrew, E. L. (2006) Changes in sediment sources following wildfire in mountainous terrain: a paired-catchment approach, British Columbia, Canada. *Water, Air, and Soil Pollution: Focus* 6, 637–645.

Owens, P. N., Blake, W. H., Giles, T. R. & Williams, N. D. (2012) Determining the effects of wildfire on sediment sources using ^{137}Cs and unsupported ^{210}Pb: the role of landscape disturbances and driving forces. *Journal of Soils and Sediments* (in press).

Paliouris, G., Taylor, H. W., Wein, R. W., Svoboda, J. & Mierzynski, B. (1995) Fire as an agent in redistributing fallout Cs-137 in the Canadian boreal forest. *Sci. Total Environ.* 160-161, 153–166.

Parsons, A. J. & Foster, I. D. L. (2011) What can we learn about soil erosion from the use of 137Cs? *Earth Science Reviews* 108, 101–113.

Petticrew, E. L., Owens, P. N. & Giles, T. (2006) Wildfire effects on the composition and quantity of suspended and gravel stored sediments. *Water, Air and Soil Pollution: Focus* 6, 647–656.

Raison, R. J. (1997) Modification of the soil environment by vegetation fires, with particular reference to nitrogen transformations: a review. *Plant and Soil* 51, 73–108.

Reneau, S. L., Katzman, D., Kuyumjian, G. A., Lavine, A. & Malmon, D. V. (2007) Sediment delivery after a wildfire. *Geology* 35, 151–154.

Rummery, T. A., Bloemendal, J., Dearing, J. A., Oldfield, F. & Thompson, R. (1979) The persistence of fire-induced magnetic oxides in soil and lake sediments. *Annales de Géophysique* 35, 103–107.

Russell, M. A., Walling, D. E. & Hodgkinson, R. A. (2001) Suspended sediment sources in two small lowland agricultural catchments in the UK. *J. Hydrol.* 252, 1–24.

Shakesby, R. A. & Doerr, S. H. (2006) Wildfire as a hydrological and geomorphological agent. *Earth-Science Reviews* 74, 269–307.

Smith, J. (1999) An introduction to the magnetic properties of natural materials. In: *Environmental Magnetism: A Practical Guide* (ed. by J. Walden, F. Oldfield & J. Smith), 5–25. Quaternary Research Association.

Smith, H. G., Sheridan, G. J., Lane, P. N. J., Nyman, P. & Haydon, S. (2011a) Wildfire effects on water quality in forest catchments: a review with implications for water supply. *J. Hydrol.* 396, 170–192.

Smith, H. G., Sheridan, G. J., Lane, P. N. J., Noske, P. & Heijnis, H. (2011b) Changes to sediment sources following wildfire in a forested upland catchment, southeastern Australia. *Hydrol. Processes* 25, 2878–2889.

Smith, H. G., Sheridan, G. J., Nyman, P., Child, D. P., Lane, P. N. J., Hotchkis, M. A. C. & Jacobsen, G. E. (2012) Quantifying sources of fine sediment supplied to post-fire debris flows using fallout radionuclide tracers. *Geomorphology* 139-140, 403–415

Taylor, R. M., McKenzie, R. M., Fordham, A. W. & Gillman, G. P. (1983) Oxide minerals. In: *Soil: An Australian Viewpoint* (Division of Soil, CSIRO), 309–335. Academic Press.

Wallbrink, P. J. & Murray, A. S. (1993) Use of fallout radionuclides as indicators of erosion processes. *Hydrol. Processes* 7, 297–304.

Wallbrink, P. J. & Murray, A. S. (1996) Determining soil loss using the inventory ratio of excess lead-210 and cesium-137. *Soil Sci. Soc. Am. J.* 60, 1201–1208.

Wallbrink, P. J., Olley, J. M. & Roddy, B. P. (1997) Quantifying the redistribution of soil and sediments within a post-harvest forest coupe near Bombala, New South Wales, Australia. CSIRO Land and Water Technical Report No. 7/97.

Wallbrink, P. J., Murray, A. S., Olley, J. M. & Olive, L. J. (1998) Determining source and transit times of suspended sediment in the Murrumbidgee River, New South Wales, Australia, using fallout Cs-137 and Pb-210. *Water Resour. Res.* 34, 879–887.

Wallbrink, P. J., Murray, A. S. & Olley, J. M. (1999) Relating suspended sediment to its original depth using fallout radionuclides. *Soil Sci. Soc. Am. J.* 63, 369–378.

Wallbrink, P. J., Roddy, B. P. & Olley, J. M. (2002) A tracer budget quantifying soil redistribution on hillslopes after forest harvesting. *Catena* 47, 179–201.

Wallbrink, P. J., Blake, W. H., Doerr, S. H., Shakesby, R. A., Humphreys, G. S. & English, P. (2005) Using tracer based sediment budgets to assess redistribution of soil and organic material after severe bushfires. In: *Sediment Budgets 2* (ed. by D. E. Walling & A. J. Horowitz), 223–230. IAHS Publ. 292, IAHS Press, Wallingford, UK.

Walling, D. E. (2005) Tracing suspended sediment sources in catchments and river systems. *Sci. Total Environ.* 344, 159–184.

Walling, D. E. & Woodward, J. C. (1992) Use of radiometric fingerprints to derive information on suspended sediment sources. In: *Erosion and Sediment Transport Monitoring Programmes in River Basins* (ed. by J. Bogen, D. E. Walling & T. J. Day), 153–164, IAHS Publ. 210, IAHS Press, Wallingford, UK.

Walling, D. E., Collins, A. L. & Stroud, R. W. (2008) Tracing suspended sediment and particulate phosphorus sources in catchments. *J. Hydrol.* 350, 274–289.

Wilkinson, S. N., Wallbrink, P. J., Hancock, G. J., Blake, W. H., Shakesby, R. A. & Doerr, S. H. (2009) Fallout radionuclide tracers identify a switch in sediment sources and transport-limited sediment yield following wildfire in a eucalypt forest. *Geomorphology* 110, 140–151.

Double trouble: the influence of wildfire and flow regulation on fine sediment accumulation in the Cotter River, Australia

MARK SOUTHWELL & MARTIN THOMS
Riverine Landscapes Research Laboratory, University of New England, Armidale, NSW 2350, Australia
mark.southwell@une.edu.au

Abstract In January 2003, the Australian Capital Territory and surrounding areas of New South Wales experienced one of the most severe wildfires in living memory. The majority of the Cotter River catchment (266 000 ha), which is a water supply region for the ACT was burnt. This study monitored the accumulation and movement of fine surficial sediment in the regulated Cotter catchment and several free flowing streams for 15 months after the fire. Significant quantities of fine surficial sediment were deposited within the channel of the Cotter River immediately following the fire. Seven months after the fire, a major rainfall event increased quantities of fine sediment by several orders of magnitude. The organic matter was significantly higher after the wildfire. Flushing flows released from the Bendora Dam removed sediment from downstream reaches causing fine surficial sediment to be preferentially eroded from riffle sections and deposited in adjacent pools. Quantities of fine surficial sediment delivered to the two unregulated streams; the Goodradigbee and Goobarragandra rivers, were much lower compared to the regulated Cotter catchment. Flows in the unregulated rivers had a greater capacity to flushing the fine material through downstream reaches because of longer duration of high flows. The results have implications for flow management and aquatic habitat in the Cotter catchment.

Key words wildfire; river regulation; fine sediment; flow management

INTRODUCTION

Wildfires have significant impacts on terrestrial and aquatic environments (Rhoades *et al.*, 2011). During severe wildfires, the majority of the ground cover is removed from catchment surfaces, leaving bare soils that are easily eroded during subsequent rainfall events. While leaf litter that falls from the burnt vegetation can help to stabilize catchment soils (Shakesby *et al.*, 2007), in most cases the extreme temperatures experienced during the wildfire reduce soil aggregate stability, increasing its susceptibility to erosion (Shakesby *et al.*, 2007). Rainfall events following wildfires can accelerate erosion from the catchment surface and increase sediment deposition in streams. This process can have significant impacts on the geomorphology, ecology and water quality of the receiving streams by increasing nutrient loads, smothering existing habitats and creating sediment slugs (Flosheim *et al.*, 1991; Lane *et al.*, 2006; Rhoades *et al.*, 2011).

Fine in-channel sediments (sand to silt sized particles) are often removed over time, following wildfires, with competent high flows flushing the excess material downstream (Flosheim *et al.*, 1991). However, in regulated rivers wildfire related sediment may not be as readily transported through the system, exacerbating the negative impacts of increased sediment loads on riverine ecosystems for longer periods of time. To promote the removal of excess fine in-channel sediments in regulated rivers, controlled dam releases, or "flushing flows" are commonly employed. The magnitude, frequency and duration of flushing flows have received much attention in the literature (see reviews by Reiser *et al.*, 1985, 1990; Kondolf & Wilcock, 1996). They are typically calculated on the basis of pre-regulation flow regime or on the requirements of stream morphology or ecology, such as flushing of fines to promote pool depth or habitat maintenance for fish species (Reiser *et al.*, 1985; Kondolf & Wilcock, 1996).

One of the most severe and destructive wildfires in the Australian Capital Territory (ACT), Australia resulted from a series of lightning strikes on 8 January 2003. The fire began after an extended drought when conditions were favourable for a catastrophic wildfire (high fuel loads, low-humidity, extremely dry soils, prevailing winds and high temperatures). The wildfires swept across native forest, farmland and into the outer suburbs of Canberra, resulting in the largest footprint in the ACT of any wildfire during the past 88 years (McLeod, 2003). The wildfire burnt 95% of the Cotter River catchment (47 000 ha), a near pristine water supply catchment, that supplies the majority of the ACT's drinking water. Although the intensity of the wildfires was patchy, the majority of this catchment experienced high to very high intensity burns with up to 15 m high flames (White *et al.*, 2006). Accordingly, the wildfire removed most of the ground cover vegetation, litter, soil organic matter, and large tracts of riparian vegetation (White *et al.* 2006) which increased the rates of hill slope runoff and soil erosion. Turbidity levels in the impoundments of the Cotter River increased due to the 2003 wildfires (White *et al.*, 2006) but little is known about post-fire sediment accumulation within the river channel. This paper documents the influence of the wildfires on the accumulation of fine sediments in the channel of the Cotter River. Sediment accumulation in the Cotter River catchment is compared to two similar river catchments to the west that were less affected by wildfires. The utility of flushing flows employed to move fine sediment through the system during the post-wildfire period is discussed.

STUDY AREA

This study was undertaken in three adjacent catchments located west of the ACT in southeastern Australia; the Cotter, Goodradigbee and Goobarragandra river catchments (Fig. 1). All three catchments have a temperate climate, with average winter and summer temperatures of approximately 6°C and 21°C, respectively. Average annual rainfall ranges from 927 mm in the east of the study area to 790 mm in the west. Rainfall event data over the study period for stations in close proximity to the three catchments are presented in Fig. 2. The Cotter River catchment has native forest covering in its headwater regions and pine plantations are located in the lower catchment. The Cotter River is regulated by three large dams: Corin Dam upstream, Bendora Dam in the mid reaches and Cotter Dam in the lower section of the river. Water is pumped out of Bendora Dam to supply drinking water to the ACT. The Goodradigbee and Goobarragandra rivers

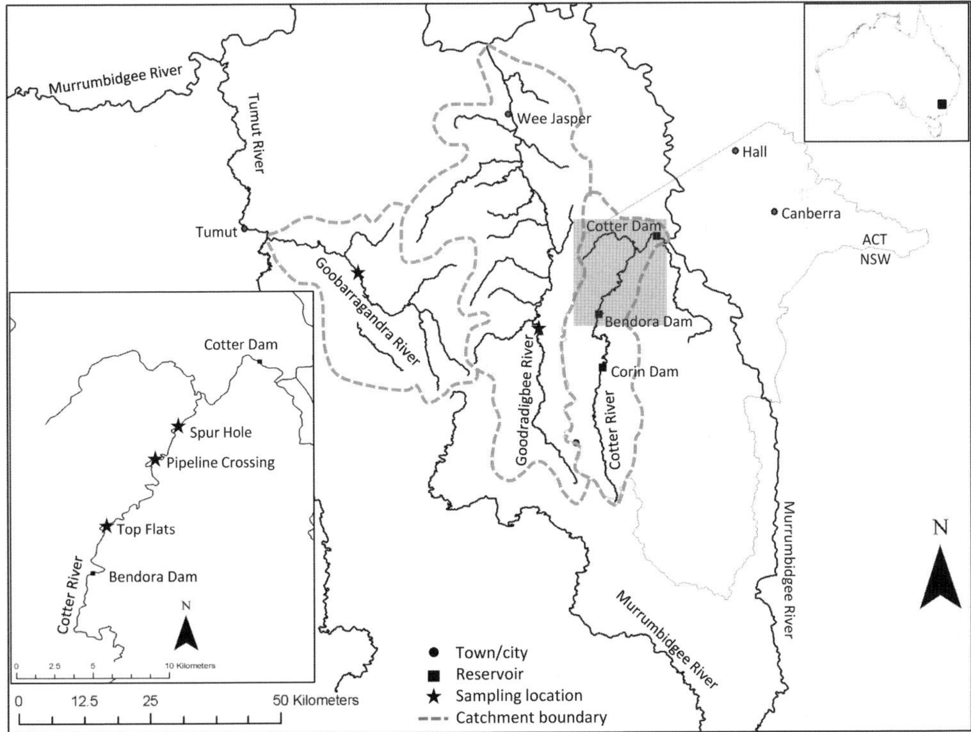

Fig. 1 Study area in southeastern Australia.

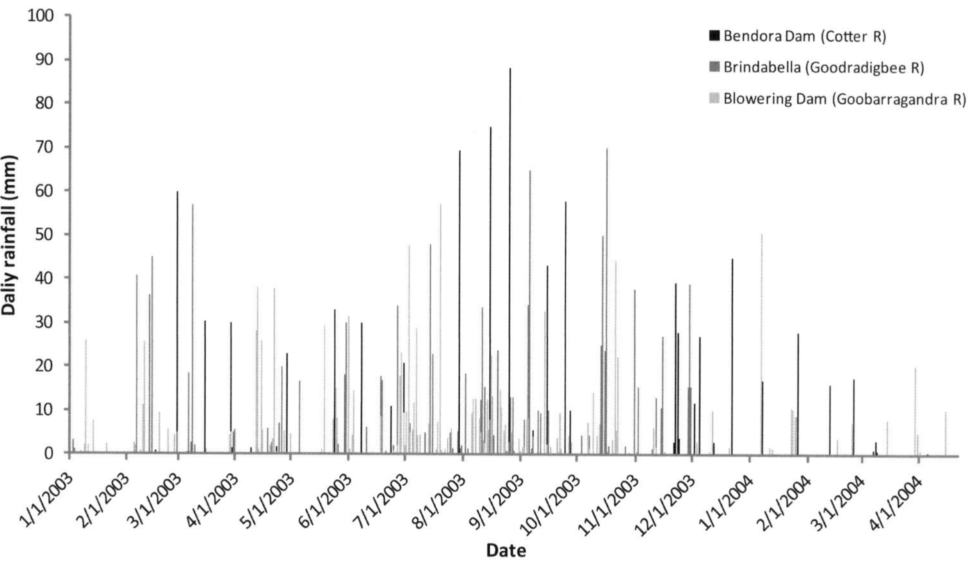

Fig. 2 Daily rainfall records for the study catchments.

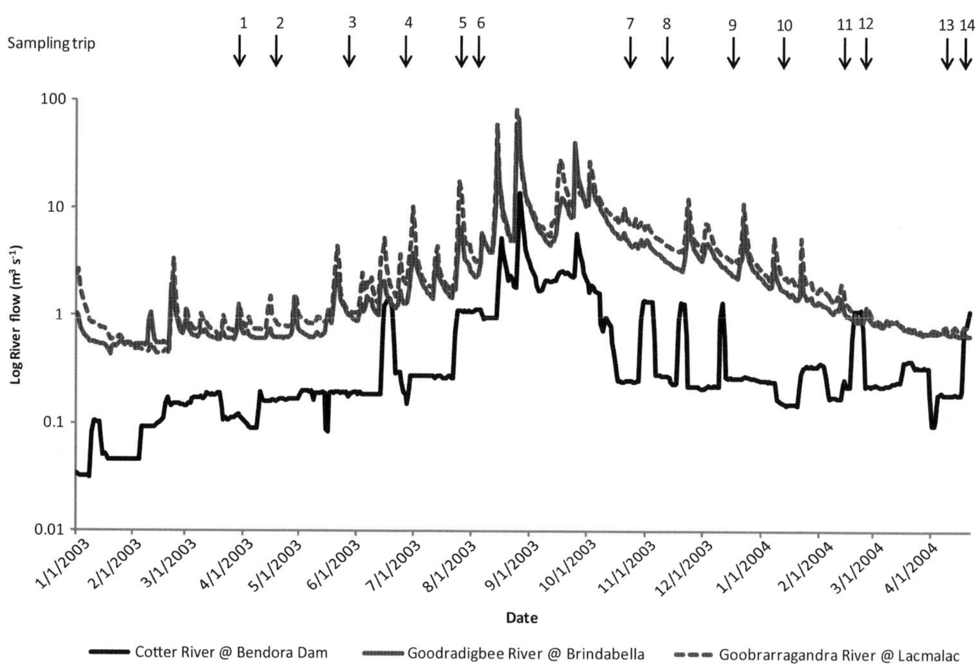

Fig. 3 Flow data. Arrows indicate the timing of monitoring trips.

are both free-flowing streams draining catchments of predominantly native forest with some pine plantation and cleared farming land. Flows in the three rivers during the study period are presented in Fig. 3.

The 2003 wildfires burned more than 95% of the Cotter and Goodradigbee river catchments, while the Goobarragandra River catchment remained un-burnt. In response to perceived increases in fine sediment accumulation within the main channel of the Cotter River, a controlled environmental flow program was conducted to maintain river habitats considered important for Macquarie Perch and Two-Spined Blackfish, which are endangered native species (Broadhurst *et al.*, 2011). The flow program included a low flow release (0.12–0.35 $m^3 s^{-1}$) followed by a riffle maintenance flow (approx 1.7 $m^3 s^{-1}$) to create flow variability and maintain spawning habitats and threatened populations of endangered fish species (Broadhurst *et al.*, 2011).

METHODS

Fine surficial sediment (Fss) samples (2 mm–32 μm) were collected at three riffle sites downstream of Bendora Dam on 18 occasions. Pre-wildfire samples were collected during 2001 and 2002 (November and December 2001, and then May and June 2002) and 14 post wildfire samples were collected from April 2003 and May 2004. Pre- and post-wildfire sampling was conducted at three individual riffle sites in both the Goodradigbee and Goobarragandra rivers. At all sites, Fss samples were collected using a modified Surber sampler with a base area of 0.09 m^2 and mesh size of 32 μm. The sampler was placed on the bed of the stream with the net and collection bottle extended in the direction of flow. The area of stream bottom within the base was disturbed by hand to entrain the fine sediment and trap it in the net and collection bottle. Five replicate samples were collected at each site on all sampling dates. Samples were dried at 40°C to a constant weight. Fine surficial sediments were not collected in pools because increased water depths meant more time consuming sampling methods would have had to been employed. The proportion of inorganic material present in each sample was determined by the procedure outlined in APHA (1997) for ash free dry mass (AFDM). A weighed sub-sample of each fine sediment sample was heated at 550°C for 2.5 hours and the loss on ignition recorded.

The downstream movement of fine sediments in the Cotter River and its impact on channel morphology was monitored by surveying paired riffle and pool cross-sections before (July 2002) and after the wildfire (September 2004). Standard channel dimensions were calculated from the survey data using the Channel program (Thoms & Ranson 2000).

RESULTS

Flows were quite different for the three rivers (Fig. 3). There were fewer pulses on the regulated Cotter River and they were considerably smaller in magnitude compared to the unregulated Goodradigbee or Goobarragandra rivers (Table 1). Mean pulse discharges were 13.33 $m^3 s^{-1}$ in the Goodradigbee River, 7.04 $m^3 s^{-1}$ in the Goobarragandra River and 4.35 $m^3 s^{-1}$ in the Cotter River. Pulses down the Goodradigbee River (defined as a flow that increased more than 1 $m^3 s^{-1}$ in consecutive days) were the longest in duration (average: 9.59 days) compared to the Cotter River (average: 8.7 days) and Goobrarragandra River (average: 7.27 days). By comparison, the four regulated flushing flows in the Cotter River had a mean pulse discharge of 0.83 $m^3 s^{-1}$ and an average duration of 8.5 days (Table 1). Base flows were smaller (average: 0.4 $m^3 s^{-1}$) and occurred for longer (87% of the time) in the Cotter River, than in the Goodradigbee River (average: 1.44 $m^3 s^{-1}$; 66% of the time) and the Goobarragandra River (average: 1.76 $m^3 s^{-1}$; 66% of the time).

Mean pre-wildfire Fss reach accumulations for the Cotter River (reach means: 8.9–15.3 g m^2) were markedly lower compared to the post period (reach means: 98.76–1392.79 g m^2). Post-wildfire, fine sediment accumulations in the Cotter River varied by over two orders of magnitude (range: 77.55–15 317 mg m^2) and this pattern was relatively consistent between sites. During relatively large rain storms that produced an extended period of flow, there was a notable increase in Fss accumulation between sampling times 6 and 7 (Fig. 4). Prior to the October–November 2003 flow event, the mean Fss for the reach was 478.90 g m^{-2}. Immediately after the flow event, mean Fss increased to 15 317.16 g m^{-2} and then decreased to a reach mean of 186.93 g m^{-2} (Fig. 4(a)).

Table 1 Flow pulse and base flow data in three rivers.

	Cotter R at Bendora Dam	Cotter River flushing flows	Goodradigbee R at Brindabella	Goobragandra R at Lacmalac
Flow pulses				
Number	7	4	17	22
Mean pulse discharge ($m^3 s^{-1}$)	4.35	0.83	13.33	7.04
Max pulse discharge ($m^3 s^{-1}$)	13.85	1.45	66.54	81.88
Min duration (days)	5	5	3	3
Max duration (days)	12	12	15	14
Mean duration (days)	8.71	8.50	9.59	7.27
Base flows				
Duration in days (%)	417 (87%)		314 (66%)	317 (66%)
Mean discharge ($m^3 s^{-1}$)	0.40		1.44	1.76

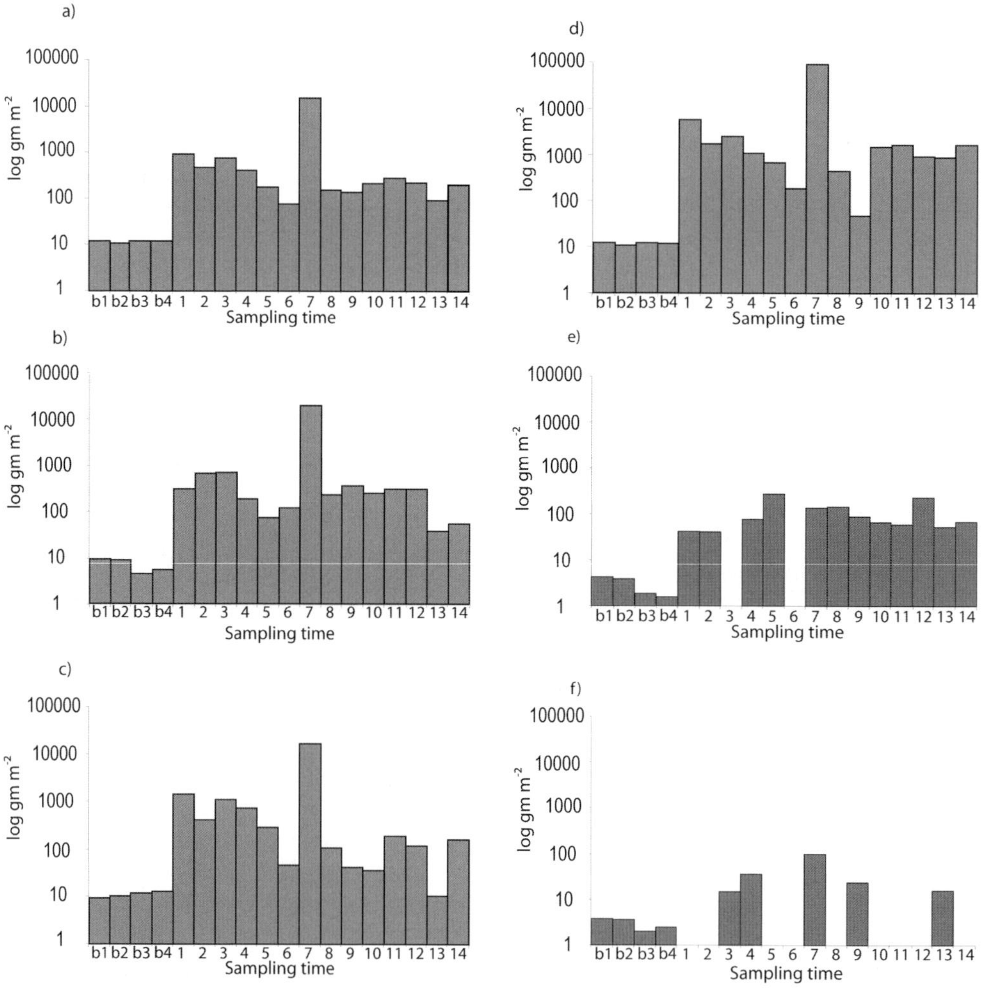

Fig. 4 Quantities of fine surficial sediment recorded during the study: (a) combined Cotter River sites (b) Top Flats site, (c) Pipeline Crossing site, (d) Spur Hole site, (e) Goodradigbee River site, and (f) Goobarragandra River site. Sampling time b1 = November 2001, b2 = December 2001, b3 = May 2002 and b4 = June 2002 and represent before fire samples. For sampling times 1–14 refer to Fig. 3.

There was considerable inter site variation in the quantity of Fss accumulation. Site means in the regulated Cotter River, ranged from 965.30 g m^{-2} at Spur Hole to 1653.66 g m^{-2} at Top Flats, while individual values ranged between 10.67 g m^{-2} at Pipeline Crossing to 19 550.78 g m^{-2} at Top Flats (Fig. 4(b)–(d)). By comparison, quantities of Fss in the Goodradigbee and Goobarragandra rivers were much lower than the Cotter River (Fig. 4(e),(f)). Means for each reach were: 1392.78 g m^{-2} for the Cotter, 107.46 g m^{-2} for the Goodradigbee and 37.40 g m^{-2} for Goodradigbee.

The majority of the Fss at the Cotter sites was organic matter. The loss on ignition ranged from 30.30 to 85.93% (Fig. 5). Overall, there was very little variation between sites (site means ranged from 66.82% at Pipeline Crossing to 67.57% at Spur Hole). There was a marked increase in LOI following the 2003 flow event at all sites; LOI increased by 32.09% at Top Flats and 45.24% at Spur Hole. The increase in Fss organic matter is presumably associated with the input of material from the catchment, much of which would have resulted from the wildfires. However, since the 2003 flow event, the proportion of organic matter in Fss has decreased at Top Flats by 8.15% and at Pipeline Crossing by 7.02%. However, it increased at Spur Hole by 14.40%. Presumably organic matter is being flushed from upstream reaches and accumulating in downstream reaches below Spur Hole.

Four flushing flows were released from Bendora Dam over the study period. The flows were short sharp releases approximately a week in duration and of magnitude ~1.4 m^3 s^{-1}. One flushing flow occurred between sampling times 7 and 8, two between sampling times 8 and 9, and one between sampling times 11 and 12 (Fig. 3). After the first flushing flow, there were marked reductions in Fss accumulation at all sites. This high flow appeared to remove the majority of sediment that was deposited after the large flow event in 2003. Successive flushing flows had much less of an influence on Fss deposits at all sites, although Spur hole did show a decline in

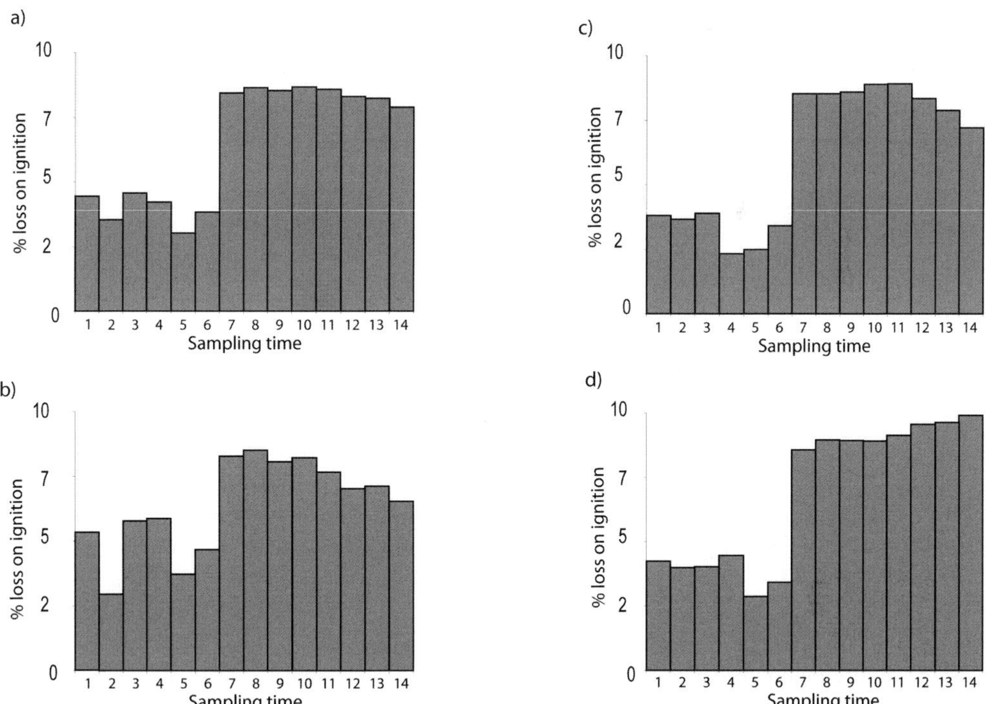

Fig. 5 Organic matter content of fine surficial sediments measured in the Cotter River study sites. (a) All sites combined, (b) Top Flats site, (c) Pipeline Crossing site, (d) Spur Hole site. For sampling times refer to Fig. 3.

sediment accumulation of around 700 g m^{-2}. The amount of sediment removed as a result of the flushing flows did not appear to be influenced by the character of the flushing event, we suggest due to the short durations of these events.

Notable changes in the channel cross-section morphology were measured after the 2003 wildfires at the Cotter River pool sites (Fig. 6). Minor change occurred at the riffle cross sections. Large changes were seen to pool cross-sectional depth and channel cross-sectional area. Channel depth decreased by 2.22 m at Top Flats, 0.88 m at Pipeline Crossing and 2.80 m at Spur Hole. Accordingly, cross-sectional areas were reduced by 57%, 36% and 55% at Top Flats, Pipeline Crossing and Spur Hole, respectively.

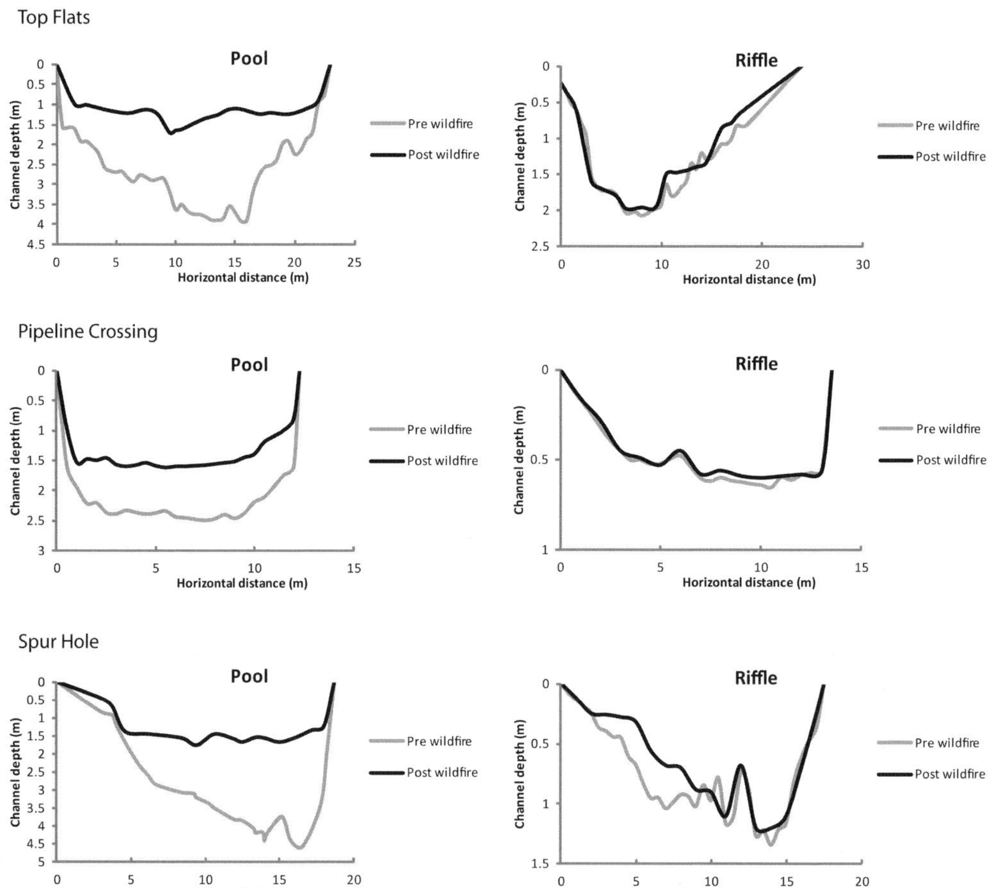

Fig. 6 Cross–sections of pools and riffles at the Cotter River sites before and after the wildfire.

DISCUSSION

Increases in the quantity of in-channel fine surficial sediment of three upland streams affected by severe wildfire highlight the combined influence of direct catchment (wildfire) and river channel (flow regulation) disturbances on riverine ecosystem processes. Increases in fine sediment accumulation in this study are comparable with other studies in the literature (Flosheim *et al.*, 1991; Minshall *et al.*, 2001; Peat *et al.*, 2005; Lane *et al.*, 2006; Rhoades *et al.*, 2011). The organic matter content of accumulated in-channel sediment increased following significant sediment inputs during the October–November 2003 high flow event. White *et al.* (2006) also noted significant increases in turbidity levels within Bendora Dam on the Cotter after several storms in early 2003 following the wildfire. These authors noted that 75% of the material transported into the dam was topsoil from the surrounding catchment rather than reworked aquatic sediments. Lane *et al.* (2006) showed that 60% of the fine sediment delivered to the East Keiwa River in southeastern Australia occurred as a result of two post-wildfire storm events.

Accumulations of fine surficial sediments in the regulated Cotter River were much larger than those recorded in either of the two unregulated catchments; even though the Goodradigbee River experienced a very similar level of burn intensity. Rainfall records for this period show that the Cotter and Goodradigbee catchments experienced rainfall events of >40mm/day throughout the sampling period (Fig. 2) that were capable of eroding the relatively bare hill slopes and transporting material to the river channels. However, the resultant river flows were quite different, highlighting the influence of flow regulation in the Cotter downstream of the dams. We suggest that the typically higher magnitude flashy flows in the unregulated Goodradigbee catchment were more capable of moving fine sediment through the system, hence the observed lower Fss accumulation. Similarly, Flosheim *et al.* (1991) observed that although the first post-wildfire storm flow deposited gravels into the channels of several unregulated streams in southern California, successive flows effectively moved sediment through the channel system. The lower Fss accumulations recorded in the Goobarragandra catchment were most likely supply-related as this catchment was not affected by the 2003 wildfires.

This study indicates the limited success of flushing flows in terms of fine surficial sediment removal. While the first flushing flow appeared to be successful in removing excess Fss accumulations from riffles deposited during several storm events in October–November 2003, successive flushing flows did not appear as effective in removing fine sediments from the sampling areas. The cross-sectional data collected in this study suggests that material may have simply been flushed from riffle sites and deposited into adjacent pools rather than being transported through the system. Flushing flows, such as the ones utilized in this study, have been advocated as a means of managing fine sediment in regulated rivers around the world (Reiser *et al.*, 1985; Reiser *et al.*, 1990; Kondolf & Wilcock, 1996; Batalla & Vericat, 2009). The magnitude and duration of these flows is very dependent on the specific objectives to be achieved (Kondolf & Wilcock, 1996). The flushing flows released downstream of Bendora Dam in the Cotter River during the study period were targeted for fine sediment removal from riffle sites, and initiation of motion calculations suggest that the flows should have been large enough to remove fine sediment from the riffle areas investigated. The apparently small influence of several of the flushing flows in removing sediment may suggest that the duration of these flows was not long enough to successfully remove significant quantities of sediment, or perhaps the extensive armoured layer on the river bed (Thoms, 2012, this volume) aided in trapping this fine sediment, restricting its removal.

Fine sediment deposited into river channels following wildfires has been shown to influence biotic communities in those systems, through changes in aspects such as water quality, temperature and habitat availability due to fine sediment accumulation (Reiser *et al.*, 1990; Minshall *et al.*, 2001; Osmundson *et al.*, 2002; Rhoades *et al.*, 2011). Indeed several studies have shown an effect of the 2003 wildfires on the macroinvertebrate and algae communities (Peat *et al.*, 2005) and benthic primary productivity (Reid *et al.*, 2012, this volume), that may be quite long lasting (Peat *et al.*, 2005). The increase in fine sediment in riffles and in particular pools of the Cotter River post-wildfire may have implications for several endangered and threatened native fish species in the river. The Cotter catchment constitutes the last remaining viable populations of both the Macquarie Perch and the Two-Spined Blackfish (Lintermans *et al.*, 2008). While both these species commonly spawn in gravel and cobble habitat within riffles (McDowall, 1996; Lintermans *et al.*, 2008), telemetry studies suggest that both species spend the majority of their time in pool habitats during a range of flow conditions (Broadhurst *et al.*, 2011). Increased sedimentation in the pools demonstrated in this study may therefore negatively impact the habitat of these endangered and threatened species.

CONCLUSION

This study demonstrates the influence of a severe wildfire on fine sediment transport and accumulation in the Cotter River in southeastern Australia. While increases in sediment were observed over a year-long period following the wildfires, the majority of it was transported during several large rainfall events in October and November of 2003. Much of the resultant sediment was removed from riffle habitats by regulated flushing flows. However, it appears to have been re-deposited in adjacent pools rather than being flushed through the system. This may have implications for the future success of several endangered and threatened fish species that frequently use pools as refuge habitats in this system.

REFERENCES

Batalla, R. J. & Vericat, D. (2009) Hydrological and sediment transport dynamics of flushing flows: Implications for management in large medditerranean rivers. *River Research and Applications* 25, 297–314.

Broadhurst, B. T., Dyer, J. G., Ebner, B. C., Thiem, J. D. & Pridmore, P. A. (2011) response of two-spined blackfish *Gadopsis bispinosus* to short-term flow fluctuations in an upland Australian stream. *Hydrobiologia* 673, 63–77.

Flosheim, J. L., Keller, E. A. & Best, D. W. (1991) Fluvial sediment transport in response to moderate storm flows following chaparral wildfire, Ventura County, southern California. *Geol. Soc. Am. Bull.* 103, 504–511.

Kondolf, G. M. & Wilcock, P. R. (1996) The flushing flow problem: Defining and evaluating objectives. *Water Resour. Res.* 32, 2589–2599.

Lane, P. N. J., Sheridan, G. J. & Noske, P. J. (2006) Changes in sediment loads and discharge from small mountain catchments following wildfire in south eastern Australia. *J. Hydrol.* 331, 495–510.

Lintermans, M., Thiem, J. D., Broadhurst, B., Ebner, B. C., Clear, R., Starrs, D., Frawley, K. & Norris, R. H. (2008) Constructed homes for threatened fishes in the Cotter River catchment: Phase 1 report. Report to ACTEW Corporation. Canberra, Institute for Applied Ecology, University of Canberra.

McDowall, R. M. (1996) *Freshwater Fishes of South-eastern Australia*. Reed Books, Chatswood, Australia.

McLeod, R. N. (2003) Inquirey into the operational response to the January 2003 bushfires in the ACT Canberra, Australian Capital Territory Government: 275.

Minshall, G. W., Robinson, C. T., Lawrence, D. E., Andrews, D. A. & Brock, J. T. (2001) Benthic macroinvertebrate assemblages in five central Idaho (USA) streams over a 10-year period following disturbance by wildfire. *Int. J. Wildland Fire* 10, 201–213.

Osmundson, D. B., Ryel, R. J., Lamarra, V. L. & Pitlick, J. (2002) Flow–sediment–biota relations: Implications for river regulation effects on native fish abundance. *Ecological Applications* 12, 1719–1739.

Peat, M., Chester, H. & Norris, R. (2005) River ecosystem response to bushfire disturbance: interaction with flow regulation. *Australian Forestry* 68, 153–161.

Reid, M. A & Thoms, M. C. (2012) Changes in benthic community structure and function in an Australian regulated upland stream following wildfire. In: *Wildfire and Water Quality: Processes, Impacts and Challenges* (Proc. conference held in Banff, Canada, June 2012). IAHS Publ. 354, IAHS Press. Wallingford, UK (this volume).

Reiser, D. W., Ramey, M. P. & Lambert, T. A. (1985) Review of flushing flow requirements in regulated streams. Pacific Gas and Electric Company. San Ramon, California. 97pp.

Reiser, D. W., Ramey, M. P. & Wesche, T. A. (1990) Flushing flows. In. *Alternatives in Regulated River Management.* (ed. by J. A. Gore & G. E. Petts), 91–135. Boca Raton, Florida, CRC Press.

Rhoades, C. C., Entwistle, D. & Butler, D. (2011) The influence of wildfire extent and severity on streamwater chemistry, sediment and temperature following the Hayman Fire, Colorado. *Int. J. Wildland Fire* 20, 430–442.

Shakesby, R. A., Wallbrink, P. J., Doerr, S. H., English, P. M., Chafer, C. J., Humphereys, G. S., Blake, W. H. & Tomkins, K. M. (2007) Distinctiveness of wildfire effects on soil erosion in south-east Autralian eukalypt forests assessed in a global context. *Forest Ecology and Management* 238, 347–364.

Thoms, M. (2012) The real issue is below the surface: Fires, fine sediment and flow regulation. In: *Wildfire and Water Quality: Processes, Impacts and Challenges* (Proc. conference held in Banff, Canada, June 2012). IAHS Publ. 354, IAHS Press. Wallingford, UK (this volume).

Thoms, M. C. & Ranson, G. (2000) Channel Program. Canberra, CRC Freshwater Ecology.

White, I., Wade, A., Worthy, M., Mueller, N., Daniell, T. & Wasson, R. (2006) The vulnerability of water supply catchments to bushfires: impact of the January 2003 wildfires on the Australian Capital territory. *Australian Journal of Water Resources* 10, 179–194.

The issue below the surface: wildfire, riverbed sediments and flow regulation

MARTIN C. THOMS
Riverine Landscapes Research Laboratory, University of New England, Armidale, NSW 2351, Australia
martin.thoms@une.edu.au

Abstract The composition and structure of upland riverbed substrates subjected to variable but multiple stressors of wildfire and flow regulation were investigated. A range of coarse–fine riverbed sediment mixtures were recorded in the upland channels draining the Brindabella Ranges, in southeast Australia and these mixtures reflected the combination of stressors studied. Significant post-wildfire increases in the accumulation of fine sediment within the riverbed substrate occurred only in association with the individual stressor of wildfire. In contrast, no change in the accumulation of matrix sediment occurred in the regulated river that was also subjected to wildfire. The presence of a well-developed surface armour layer – a feature of gravel-bed regulated rivers – prevented the infilling of these riverbed substrates post-wildfire. Further study of the combined impacts of different stressors on the composition and structure of upland gravel-bed rivers will contribute to an improved understanding of their recovery from wildfire.

Key words sub-surface riverbed sediment; wildfire; fine sediment; flow regulation

INTRODUCTION

The composition and structure of riverbed substrates is important for numerous ecological, engineering and geomorphological concerns. Their evolution, sedimentation and the influences of direct and indirect drivers were reported as one of nine major scientific hydrological challenges that require further study (UNESCO, 1984). The natural character of gravel-bed substrates is complex, consisting of a wide range of sediment textures and an array of coarse-fine sediment mixtures in natural river channels. Riverbed substrates can acquire quasi-uniform characteristics that reflect prevailing hydrological and sedimentological regimes operating at a range of scales. Because of the inherent heterogeneity and spatial variability of riverbed substrates, many previous studies have tended to focus on the character and behaviour of surface layer sediments only, ignoring perhaps the relatively more important sub-surface sediment. Studies examining the character of sub-surface riverbed sediments have demonstrated their spatial and temporal complexity often in association with a range of catchment conditions (Thoms, 1992).

Riverine ecosystems are frequently subjected to multiple stressors and interactions between individual stressors. Either through positive or negative feedback loops, these stressors can enhance or diminish expected individual impacts. Despite the ubiquitous nature of multiple stressors, their impact on riverine ecosystems has received limited attention (Thoms, 2007). Recently Ormerod (2010) hypothesised that antagonistic and synergistic responses to multiple stressors may range from simple additive impacts to those that are increasingly complex. Catchment wildfires and the regulation of flows through dam construction are amongst the most significant stressors on riverine ecosystems (Stanley *et al.*, 2010), especially in headwater systems. The impact of wildfires on catchment runoff, sediment yields and resultant changes in channel morphologies, water and sediment quality, biological communities and instream ecological productivity are well documented (Minshall, 2003). In particular, large quantities of fine sediment can accumulate within those river channels whose catchments have been extensively burnt (Beschta *et al.*, 2004). Recoveries of riverine ecosystems following wildfires can be slow in river systems that are also subject to additional stressors such as the grazing of livestock and water resource development (Beschta *et al.*, 2004). Flow regulation directly affects sediment and water regimes, which are key drivers of the structure and composition of riverbed sediments and riverine ecosystems, and have the potential to influence recovery following wildfire disturbance. Sediment movement is a function of the river capacity and competence, as well as sediment availability in the catchment. Either factor can limit sediment transport rates and therefore influence the character of riverbed sediments. Despite increasing evidence of the compound nature and negative effects of multiple stressors influencing rates of recovery in riverine ecosystems, relatively few studies have examined the impact of wildfire in riverine ecosystems regulated by dams.

In January 2003, wildfires burned over 500 000 ha of the Southeast Highlands of Australia, including large sections of the Brindabella Ranges within the Australian Capital Territory. Some of the impacted areas are key source water catchments for Canberra, the nation's capital. Riverine ecosystems that drain these catchments also support several endangered native fish species that rely on riverbed substrates for various stages of their life-cycles. In an effort to reduce the impact of the 2003 wildfires, the local water authority implemented a series of flow releases to remove excess fine sediments that accumulated in the river channel of the main regulated system of the region. This provided an opportunity to investigate the multiple impacts of wildfire and flow regulation on the sedimentological character of the riverbed substrates. The research questions addressed in this study are: (1) what is the effect of multiple stressors (wildfire and flow regulation) on the composition and structure of riverbed substrates; and, (2) do reservoir releases enhance the recovery of riverbed substrates subjected to these multiple stressors.

STUDY AREA

This study was undertaken in three adjacent upland forested catchments within the Brindabella Range of SE Australia: the Cotter, Goodradigbee and Goobarragandra rivers (Fig. 1). All three catchments have a temperate climate with a long-term average regional rainfall (1954–2008) of 930 mm, most of which occurs during August–October. Granites are prevalent on the ridges with Ordovician sediments (shales, sandstones and clays) on the slopes. Land use over most of the three catchments is National Park or Nature Reserve, with limited commercial forestry in the lower Cotter catchment (<8% of total catchment) and some rural grazing/cultivation along the mid to lower sections of the Goodradigbee and Goobarragandra rivers. The study catchments are typical mountain rivers (Wohl, 2010) flowing across highly constrained valleys with characteristic cobble/gravel bed riverbed sediments. The Cotter River has a catchment area of 482 km^2 and is regulated by three dams (Corin, Bendora and Cotter), whereas the unregulated Goodradigbee and Goobarragandra rivers have catchment areas of 890 km^2 and 673 km^2, respectively. Environmental flow releases designed to minimize the impact of the dams and mimic the natural flow regime occur in the Cotter River.

In January 2003, lightning started wildfires that burned the majority (>95%) of the Cotter and Goodradigbee. The Goobarragandra catchment was not burned. In an attempt to manage significant increases in the accumulation of fine sediment within the main channel of the Cotter River post wildfire (Southwell & Thoms, 2012), the local water authority conducted a series of "flushing flows" to remove "excess" sediment from the channel and maintain river habitats considered to be important for two endangered native fish species, the Macquarie Perch and Two-Spined Blackfish, (Broadhurst *et al.*, 2011). Both species utilize the riverbed during parts of their life cycle and the nature of the substrate is important for their recruitment. The release of flushing flow includes a pre-release of 0.12–0.35 $m^3 s^{-1}$ that increased to 1.7 $m^3 s^{-1}$; the latter release is designed for riffle maintenance.

STUDY DESIGN

A modified BACI study design was employed in this study. Riverbed sediments were collected pre-wildfire during December 2001, and then on two occasions post-wildfire. The first post-wildfire sampling period was several weeks after significant regional rainfall (June 2002) and the second, several weeks after the first flushing flow in the Cotter River. This sampling period occurred after several floods in the Goodradigbee and Goobarragandra rivers (May 2004). Bed material samples were collected by the freeze coring method (Thoms, 1992). This technique efficiently collects fine sediment, enables undisturbed volumetric samples to be obtained and

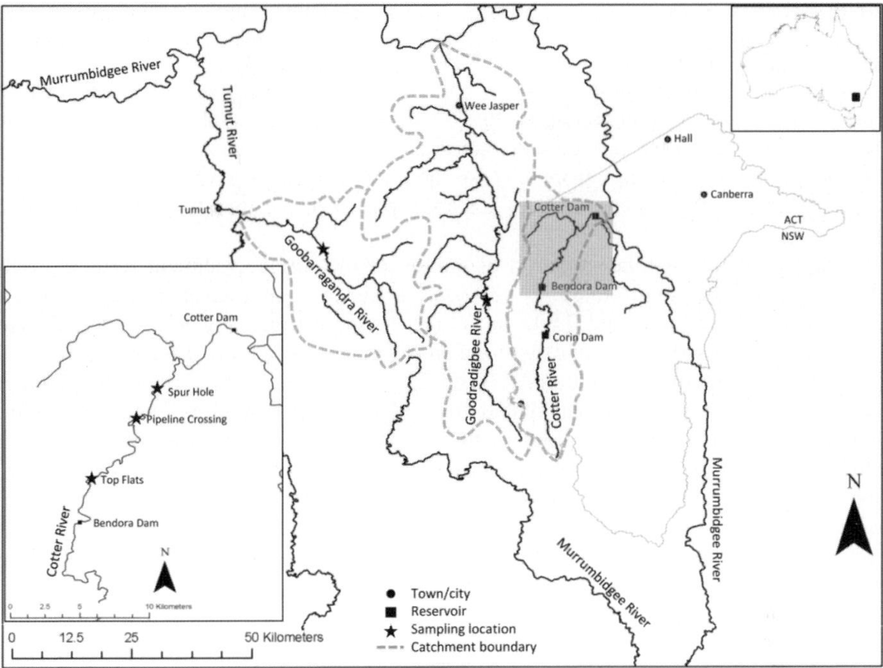

Fig. 1 Study catchments and sample locations.

allows the composition and depositional history of riverbed substrates to be examined. The surface/armour layer was removed first, before the freeze core samples were collected, because this study was primarily concerned with the accumulation of fine sediment within the riverbed. Five freeze-core samples, weighing approximately 10 kg each, were removed from three riffle sites in the Cotter River and two riffle sites in the Goodradigbee and Goobarragandra rivers on each sampling occasion (n = 105). This sampling effort ensured the percent weight of the largest particle was less than five percent of the total sample weight (Mosley & Tinsdale, 1985) and to further reduce the influence of the largest particle, samples were combined to produce a composite sample for each site. Samples were dried at room temperature and then sieved to obtain graphic statistical measures, with results expressed in phi (Ø) units.

The resultant grain-size data were analysed using the multivariate statistical procedure (Forrest & Clark, 1989). Entropy analysis is a nonparametric clustering technique where individual sediment samples are represented by the entire grain-size distribution and the analysis classifies samples into groups sharing common grain-size distribution characteristics. The optimum number of groups, or classes, is obtained when the between-class entropy similarity increases at a significantly decreasing rate with the addition of more classes (Forrest & Clark, 1989). Grouping of samples with similar sediment character based on the entire grain-size distribution offers advantages over methods that use only summary statistics of individual samples. In addition, a series of standard textural statistical measures were calculated, with between site and sampling period differences tested via an Analysis of Variance (ANOVA).

Surface sediment sampling was also undertaken at each site during December 2001. The Wolman random walk method was employed with the intermediate "b" axis of 250 surface sediment particles counted at each site. Statistical differences between the grain-size distributions of each site were determined via the Kolomogorov-Smirnov test. The initiation of motion of surface sediments at each site was calculated using the Schoklitsch bed load transport equation (Bathurst *et al.*, 1987). In addition, an Armouring Index (AI) and the Relative Bed Stability Index (RBS) for all sites were calculated via equations given by Kaufmann (1999).

RESULTS

Surface sediments at all sites are comprised of particles ranging from 0 to –8 ϕ: very coarse sand to cobble material on the Wentworth size scale. Overall, surface sediments are dominated by cobble-sized particles (–6 to –8 ϕ) comprising 63 to 77% by weight of the total sample mass. There was no statistical difference in the surface sediment grain-size distributions between sites at the 0.05% level. Armour index values range from 1.97 for the Goobarragandra River to 3.27 at Spur Hole in the Cotter River; typical values of 1.2 to 2.8 are reported in the literature (e.g. Bathurst *et al.*, 1987).

Discharges required for the initiation of motion of the surface sediment (critical discharge) varied between sites; ranging from 85.38–381.80 $m^3 s^{-1}$ for the Cotter River sites to 62.83 $m^3 s^{-1}$ for the Goodradigbee River sites and 75.42 $m^3 s^{-1}$ for the Goobarragandra River sites. Presumably these differences are associated with differences in stream power in each of the rivers.

Combining the initiation of motion data with flow data for each of the rivers, the frequency of surface sediment movement was calculated for 1987–2002 (pre-wildfire) and 2003–mid 2004 (post-wildfire) periods (Table 1). Flows in the regulated Cotter River were not of a sufficient magnitude to initiate movement of the coarser surface sediment during either the pre- or post-wildfire period. In contrast, the initiation of motion in the unregulated Goodradigbee and Goobarragandra rivers was exceeded 1.54 and 1.93% of the time in the post-wildfire period, respectively. During the pre-wildfire period, the initiation of motion was exceeded 3.2% of the time at sites in the Goodradigbee River and 2.81% of the time at sites in the Goobarragandra River (Table 1). This variation in the relative stability of the surface riverbed sediment is confirmed by the RBS values. RBS values are an order of magnitude higher for the Cotter River sites compared to those in the Goodradigbee and Goobarragandra rivers. RBS values >1 indicate a stable riverbed, while those <1 indicate an unstable riverbed (Jowett, 1989). By comparison, flows were of a sufficient magnitude to move finer surface sediments at all sites and the frequency of occurrence ranged from 14.67% of the time (75 days) at Pipeline to 55.55% of the time (287 days) at Site 2 in the Goobarragandra River during the post-wildfire period.

Composite grain-size histograms for the three rivers show the sub-surface sediment to be bimodal (Fig. 2). There is a coarser mode (the "framework") with particles between –10 and –4ϕ and a secondary mode of finer sediments (the matrix) with particles between –2 and +4ϕ. These two modes can be considered separate populations. The resolution of the two individual modes is commonly based on the presence of a saddle frequency, in this case located at –3ϕ. This bimodal grain-size distribution is a feature of all sites in the study area and is indicative of high-energy upland gravel-bed rivers. As a first approximation the coarser framework deposit, i.e. sediment larger than –4ϕ, represents the traction load of the channel while the finer matrix sediment, i.e. that finer than –2 ϕ, is characteristic of a saltating suspended load. Commonly grain-size analyses are based on the assumption of a Gaussian distribution, therefore Carling & Reader (1982) suggest that with bimodal sediment populations, each mode should be analysed separately.

Table 1 Frequency of coarse surface sediment movement and riverbed stability.

Site	No. of days of sediment movement		% occurrence		RBS
	Pre-wildfire	Post-wildfire	Pre-wildfire	Post-wildfire	
Cotter River – multiple stressors of wildfire and flow regulation					
Burkes Creek	0	0	0	0	12.5
Pipeline	0	0	0	0	10.2
Spur Hole	0	0	0	0	7.3
Goodradigbee River – singe stressor of wildfire					
Site 1	187	10	3.2	1.93	0.98
Site 2	187	10	3.2	1.93	1.02
Goobarragandra River – no stressor					
Site 1	164	8	2.81	1.54	1.01
Site 2	164	8	2.81	1.54	0.99

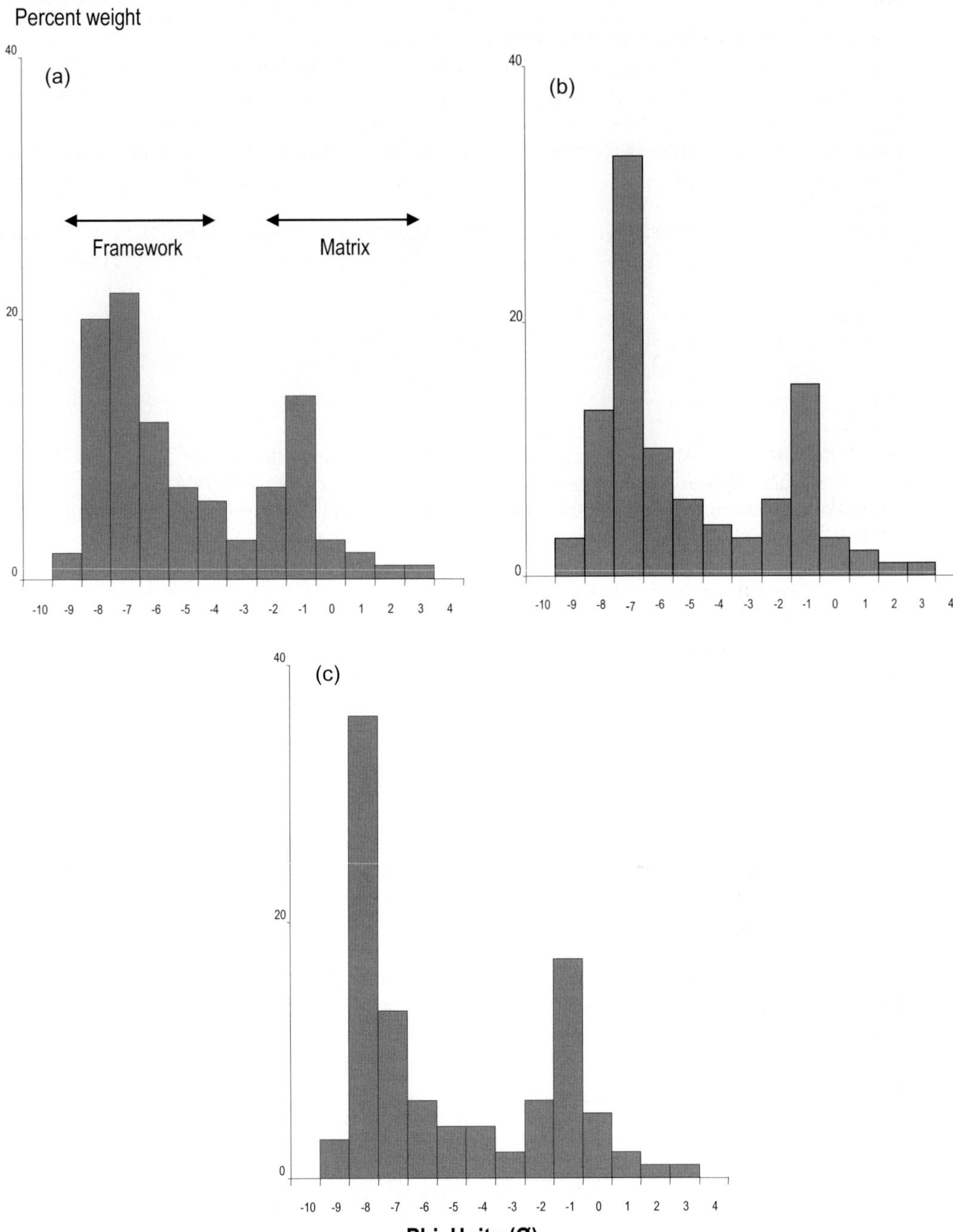

Fig. 2 Composite grain-size histograms for sub-surface riverbed sediments for: (a) Cotter River; (b) Goodradigbee River; and (c) Goobarragandra River.

Table 2 The contribution of matrix sediments to the sub-surface sediment in the study area. The average and range of the percentage weight of matrix are given for the three sampling periods are provided with the ratio of average pre- to post-wildfire matrix contribution, are given.

River	Pre-wildfire	Post-wildfire I	Post-wildfire II
Cotter			
Average	25.97%	26.12%	25.23%
Range	22.88–31.87%	23.11–31.90%	23.51–30.75%
Ratio		1	0.91
Goodradigbee			
Average	18.23%	51.23%	36.21%
Range	15.99–21.21%	42.52–59.68%	34.15–39.51%
Ratio		2.81	1.97
Goobarragandra			
Average	16.87%	17.23%	17.01%
Range	15.37–17.93%	16.21–19.18%	16.88–19.17%
Ratio		1.02	1

There was no statistical difference in the mean grain size of the coarser framework mode of the sub-surface sediments between sites or sampling periods (ANOVA: $F_{(0.05),3,21} = 5.3$). Framework mean grain sizes for the Cotter, Goodradigbee and Goobarragandra rivers ranged from –6.89 to –7.21 ϕ, –6.90 to –7.31 ϕ and –6.85 to –7.19 ϕ, respectively.

The range in the contribution of the matrix component to riverbed sub-surface sediment between different sites and sampling periods was a dominant feature of the individual grain-size histograms (Table 2). The contribution of matrix sediment varied most in the Goodradigbee River over the three sampling occasions (15.99 to 59.68% by weight) compared to the Cotter (22.88 to 31.90 % by weight) and Goobarragandra Rivers (15.37 to 19.18% by weight). There was a significant difference in the percent weight of the matrix sediment between the three rivers (ANOVA: $F_{(0.05),3,21} = 25.6$). However, there was no significant difference in the percentage weight of the matrix sediment between sites or sampling period in the Cotter River (ANOVA: $F_{(0.05),7,21} = 34.2$). Thus, wildfire had no effect on the accumulation of finer matrix sediments within the riverbed substratum of the Cotter River. Similarly, there was no significant difference (ANOVA: $F_{(0.05),2,6} = 2.6$) between sites or sampling period in the Goobarragandra River. In contrast, there was a significant difference (ANOVA: $F_{(0.05),2,6} = 14.1$) between sampling events in the Goodradigbee River. Although the percentage weight of matrix sediment for the Goodradigbee sites was comparable to that recorded at the Goobarrangandra sites for the pre-wildfire sampling period (an average of 18.23 and 16.87% by weight for the Goodradigbee and Goobarrangandra, respectively), there were marked differences between the two rivers in the two post-wildfire sampling periods. Immediately following the wildfires, the contribution of the matrix component increased on average by 181% (18.23 to 51.23% by weight) in the Goodradigbee, while the difference between the pre-wildfire and sampling period 2 post-wildfire the difference was only 98.63% (18.23 to 33.21% by weight).

Results of the entropy analysis confirm differences in the overall grain-size composition of the sub-surface sediment between some rivers and sampling periods. Entropy analysis identified three distinct sub-surface sediment texture groups accounting for 83.2% of the total variation between individual samples. Group 1 contained all the Cotter River samples, while Group 2 contained all the Goobarragandra samples plus the pre-wildfire Goodradigbee samples, and Group 3 was comprised of post-wildfire samples from the Goodradigbee. These differences in the overall grain-size composition were a result of variations in the contribution of the finer matrix sediment.

DISCUSSION

The purpose of this paper was to detail the sub-surface sedimentological character of three upland rivers subjected to varying multiple stressors and to assess the recovery of riverbed substrates whose catchment have been subjected to wildfire. The substrate of gravel-bed rivers can be separated into four broad categories: open-work gravels, matrix-filled contact or framework dominated gravels, framework dilated gravels, and matrix dominated, on the basis of framework packing and matrix contribution. Where the matrix forms less than 25% of the total sediment weight the substrate is termed open-work gravels; when the matrix is 25–32% of the sediment weight, the substrate is "framework dominated" in which the clasts are in tangential contact with one another (Carling & Reader, 1982) forming a stable self-supporting structure but interstitial spaces between framework clasts are occupied by the finer matrix sediment. Further contributions of matrix sediment theoretically reduce the stability and inter-clast contact of the coarser framework sediments. A range of riverbed substrates was recorded in the three rivers. Framework dominated sub-surface sediments were present in the Cotter River, while open-work gravels characterise the Goobarragandra and the pre-wildfire Goodradigbee sediments. By contrast the post-wildfire sub-surface sediments of the Goodradigbee River are framework dilated and matrix dominated.

Large quantities of fine matrix sediments can be stored within coarse riverbed substrates. In this study, up to 59% by weight of fine matrix sediment was recorded within the substrate of the Goodradigbee River, in comparison with a maximum of 31.91% in the Cotter River and 19.17% in the Goobarragandra River. The supply of finer sediment to the riverbed surface is the primary factor governing the accumulation of matrix sediment within coarser riverbed substrates. The ingress of matrix sediment into a framework substrate can be rapid, even at low concentrations (Carling, 1984). In this study, rates of fine sediment deposition increased by three orders of magnitude within those river channels whose catchments were subjected to wildfire (Southwell & Thoms, 2012). This was prevalent following the first major rainfall event post-wildfire and for over 18 months following this catchment disturbance. Increased accumulations of fine matrix sediments within coarse riverbed substrates have also been reported to occur in association with a range of catchment stressors such as urban development (Thoms, 1987), dam construction (Davey *et al.*, 1987) and forestry activities (Eibse, 1983).

The results of this study demonstrate variations exist in the response of riverbed substrate composition to different individual stressors and stressor combinations. Goodradigbee River sites, subjected to the single stressor of wildfire, experienced the greatest change in the composition of the riverbed substrate. Average change ratios of 2.81 and 1.97 for the percentage weight of matrix sediment were recorded for the two post-wildfire sampling periods in the Goodradigbee River compared to those in the Cotter River – sites subjected to the multiple stressors of wildfire and flow regulation (1 and 1.01), and those in the Goobarragandra River which experienced no stressor (1.01 and 1) (cf. Table 2). It appears there was no additive response from the combined stressors of flow regulation and wildfire on the composition of the riverbed substrate in the Cotter River. Rather, the single wildfire stressor had a greater impact on the changing the textural composition of the riverbed substrate (Table 2). However, the marked reduction in the percentage weight of matrix sediment between the two post-wildfire sampling periods in the Goodradigbee River (average percent reduced from 51.23 to 36.21%) suggests the impact may be transient. Competent river flows, as experienced during floods that result in movement of coarser framework sediments and turnover of the riverbed substrate, will flush finer sediment from the sub-surface substrate (Gomez, 1987). In contrast, the pre-wildfire conditions show a higher matrix component within the riverbed substrate of the regulated Cotter River compared to the other two rivers, and this did not change significantly post-wildfire (Table 2). This suggests a more synergistic response to the combined stressors of wildfire and flow regulation.

The presence of fine sediments within coarse-bed river deposits has been inferred to result from the processes of: simultaneous deposition (Frazer, 1935), particle overpassing (Allen, 1983), or secondary infiltration or ingress (Einstein, 1968). Regardless of the actual mechanism of accumulation, the vertical structure of riverbed substrates influences the residence of matrix sediments within the substrate overtime. The vertical structure of gravel-bed substrates is rarely uniform, with the presence of a coarse surface layer being a feature of heterogeneous gravel-bed sediments. Generically termed the armour layer, it has also been referred to as a pavement or censored layer, depending on its relative mobility. Once formed, the armour layer protects riverbed turnover and disturbance of the sub-surface deposit during floods. During sediment transport where the riverbed is mobilised, finer sediment is flushed out of the substrate effectively, increasing its sub-surface porosity. A well-developed armour layer is a feature of the regulated Cotter River, resulting in highly stable riverbed and framework dominated sub-surface gravels where all interstitial spaces are occupied by matrix sediment. Since dam construction in the Cotter River there has been no bed load movement and therefore no disturbance of the sub-surface sediments. In the unregulated Goodradigbee and Goobarragandra rivers by comparison, there is no well-developed armour layer and as a consequence riverbed sediments are more frequently disturbed. This not only cleanses the sub-surface gravels, but also allows excess matrix accumulation to occur during periods of elevated fine sediment supply, as happens post-wildfire.

Programmes of predetermined flow releases, downstream of dams, for a given duration are frequently employed to meet a variety of management goals in regulated rivers. The maintenance of some "desired" channel characteristic is a common goal and these flow releases are termed "channel maintenance flows" or, more commonly, "flushing flows" for the effect of removing (flushing) fine sediments from riverbed gravels. Following wildfires in the study region, a programme of flushing flows was implemented in the Cotter River to remove 'excess' fine sediments, that had accumulated as a result of the wildfires and improve habitat conditions for two endangered native fish species. The results of this study have demonstrated that this programme of flushing flows had no measureable effect on the sediment composition of the sub-surface riverbed. This presumably was because of the presence of a well-developed surface armour layer that prevented sediment movement bed turnover and the flushing of fine sediments from within the substrate.

Acknowledgements Many undergraduate students from the University of Canberra's Catchment Science Unit assisted with the collection of field data – to them a debt of gratitude is owed. The majority of the research undertaken in the Cotter Catchment was undertaken in collaboration with Prof. Richard Norris; I enjoyed the journey Richard and may you rest in peace. Comments on an earlier draft of this manuscript by Dr Melissa Parsons were appreciated.

REFERENCES

Allen, J. R. L. (1983) Gravel overpassing on humpback bars supplied with mixed sediment: Examples from the Old Red Sandstone, southern Britain. *Sedimentology* 30, 285–294.

Bathurst, J. C., Thorne, C. R. & Hey, R. D. (eds) (1987) *Gravel Bed Rivers: Fluvial Processes, Engineering and Management.* Wiley, Chichester, UK.

Beschta, R. L., Rhodes, J. J., Kauffman, J. B., Griesswell, R. E., Minshall, G. W., Karr, J. R., Perry, D. A., Hauer, E. R. & Frissell, C. A. (2004) Postfire management on forested public lands of the western United States. *Conservation Biology* 18(4), 957–967.

Broadhurst, B. T., Dyer, J. G., Ebner, B. C., Thiem, J. D. & Pridmore, P. A. (2011). Response of two-spined blackfish *Gadopsis bispinosus* to short-term flow fluctuations in an upland Australian stream. *Hydrobiologia* 673, 63–77.

Carling, P. A. (1984) Deposition of fine and coarse sand in an open-work gravel bed. *Can. J. Fisheries and Aquatic Sciences* 41, 263–270.

Carling, P. A. & Reader, N. A. (1982) Structure, composition and bulk properties of upland stream gravels. *Earth Surface Processes and Landforms* 7, 349–365.

Davey. G. W, Doegm T. J & Blyth, J. D. (1987) Changes in benthic sediment in the Thomson River, Victoria, during construction of the Thomson Dam. *Regulated Rivers: Research and Management, Research and Management* 1.

Ebise, S., Aizaki, M., Otsubo, K. & Muraoka, K. (1983) Estimation of river sediments as pollutants loadings. *National Institute for Environmental Studies, Japan* 6, 93–103.

Einstein, H. A. (1968) Deposition of suspended particles in a gravel bed. *J. Hydraulics Div. ASCE* 94, 1197–1205.

Frazer, H. J. (1935) Experimental study of the porosity and permeability of clastic sediments. *J. Geology* 43, 910–1010.

Forrest, J. & Clark, N. (1989) Characterising grain-size distributions: evaluation of a new approach using multivariate extension of entropy analysis. *Sedimentology* 36, 711–722.

Gomez, B. (1983) Temporal variations in bedload transport rates: the effect of progressive armouring. *Earth Surface Processes and Landforms* 8, 41–54.

Kaufmann, P. R., Faustini, J. M., Larsen, D. P. & Shirazi, M. A. (1999) A roughness corrected index of relative bed stability for regional stream surveys. *Geomorphology* 99, 150–170.

Minshall, G. W. (2003) Responses of stream benthic macroinvertebrates to fire. *Forest Ecology and Management* 178(1-2), 155–161.

Mosley, M. P. & Tindale. R. S. (1992) Sediment variability and bed material sampling in gravel bed rivers. *Earth Surface Processes and Landforms* 10, 465–482.

Ormerod, S. J., Dobson, M., Hildrew, A. G. & Townsend, C. R. (2010) Multiple stressors in freswhater ecosystems. *Freshwater Biology* 55, 1–4.

Reid, M. A., Thoms, M. C. & Southwell, M. (2012) Changes in benthic community structure and function in an Australian regulated upland stream following wildfire. In: *Wildfire and Water Quality: Processes, Impacts and Challenges* (Proc. conference held in Banff, Canada, June 2012). IAHS Publ. 354, IAHS Press. Wallingford, UK (this volume).

Southwell, M. & Thoms, M. C. (2012) Double trouble: the influence of wildfire and flow regulation on fine sediment accumulation in the Cotter River, Australia. In: *Wildfire and Water Quality: Processes, Impacts and Challenges* (Proc. conference held in Banff, Canada, June 2012). IAHS Publ. 354, IAHS Press. Wallingford, UK (this volume).

Stanley, E. H., Powers, S. M. & Lottig, N. R. (2010) The evolving legacy of disturnace in stream ecology; concepts, contributions and coming challenges. *J. North American Benthological Society* 29, 67–83.

Thoms, M. C. (1987) Channel sedimentation in the urbanised River Tame, UK. *Regulated Rivers: Research and Management* 1, 229–246.

Thoms, M. C. (1992) A comparison of grab- and freeze-sampling techniques in the collection of gravel-bed sediment. *Sedimentary Geology* 78, 191–200.

Thoms, M. C (2007) The distribution of heavy metals in a highly regulated river: the River Murray, Australia. In: Water Quality and Sediment Behaviour of the Future (ed. by B. W. Webb & D. de Boer), 145–153. IAHS Publ. 314. IAHS Press, Wallingford, UK.

UNESCO (1984) International Hydrological Challenges. Ref. NS/188, Annex II, Paris.

Wohl, E. E. (2010) *Mountain Rivers Revisited.* Water Resources Monograph 19, American Geophyiscal Union.

Reducing wildfire risk in water supply catchments using payments for ecosystem services

ASHLEY A. WEBB
Forests NSW, PO Box 4019, Coffs Harbour Jetty, NSW 2450, Australia
ashleyw@sf.nsw.gov.au

Abstract In New South Wales (NSW), Australia, local combinations of sclerophyllous vegetation dominated by *Eucalyptus* species, rapid fuel accumulation, terrain and weather can result in a high probability of uncontrollable wildfires. For example, in December 2001 and January 2002 the "Black Christmas" bushfires burned 733 342 ha of forest, including 225 000 ha within the Sydney water supply catchments. Evidence from several studies in NSW indicates that the effects of wildfires pose an unacceptable risk to water supplies. A parliamentary inquiry into the Sydney fires has resulted in a greater focus on hazard reduction. As forests in NSW occur on a mixture of land tenures, legal, institutional and economic barriers limit the effectiveness of efforts to reduce wildfire risk. This paper introduces the concept of payments for ecosystem services (PES) and discusses successful schemes that have recently been implemented to reduce the risk of wildfires in catchments supplying water to the cities of Santa Fe, New Mexico and Denver, Colorado in the USA.

Key words payments for ecosystem services, PES; hazard reduction; water supply catchments; forest management

INTRODUCTION

One-third of the Australian population lives in the state of New South Wales (NSW) where forests cover 26.59 million ha or approximately 33% of the land area (ABARES, 2011). These forests occur on various land tenures including leasehold land (35%), private land (32%), National Parks and conservation reserves (17%), multiple-use State forests (9%) and Crown land (4%). In the most populated regions along the east coast and ranges, native *Eucalyptus* forests are largely relied upon to provide a continuous supply of high quality water for domestic, agricultural and industrial use (Webb, 2012a). This reliance on forests for water supplies in the Australian context is, however, potentially fraught as wildfires are a recurring feature of the landscape (Singh *et al.*, 1981). For example, the region surrounding Sydney, the largest city in NSW, has a long history of wildfires with significant recent events recorded in 1952, 1957, 1968, 1977, 1982, 1993/4 and 2001/2 (Dragovich & Morris, 2002a). North of Sydney, the fire season, when serious fires are likely to occur, runs from September to December. Around Sydney, and to the south, the fire season runs from October to January. However, occasional serious fires have been known to occur in February and March (Cheney, 1995). It is this part of NSW that experiences the most severe fire weather; high pressure systems in the Tasman Sea and low pressure systems in the Southern Ocean can create high pressure gradients that direct hot air from the arid interior of the continent to the southeastern part of Australia (Cheney, 1995). Essentially all of the vegetation, mostly dominated by *Eucalyptus* species, is susceptible to fire.

During December 1993 and January 1994, over 800 fires burnt >800 000 ha of forest in eastern NSW, destroying 287 premises and causing four deaths (Cheney, 1995). Similarly, in December 2001 and January 2002, the large "Black Christmas" wildfires burnt 733 342 ha of forest. These fires cost the NSW government A$217 million to suppress, cost a further A$80 million in insurance claims, destroyed 121 premises and caused injuries to 50 people. Despite the widespread nature of these severe fires and their proximity to high population urban areas, zero deaths were recorded (Coghlan, 2004). A salient point, however, is that the 2001/2 fires burnt 225 000 ha of forests within the source catchments supplying Sydney's water (Chafer *et al.*, 2004).

A wide range of soil erosion and water quality impacts due to wildfires in eastern NSW has been reported over recent decades. Blong *et al.* (1982), Atkinson (1984), Zierholtz *et al.* (1995)

and Dragovich & Morris (2002a,b) have all documented increases in hillslope soil erosion following bushfires in the greater Sydney region. However, as Prosser & Williams (1998) indicated, the risk of accelerated soil erosion and delivery to stream networks is highly dependent upon the timing and magnitude of rainfall events experienced in the post-fire period. On the south coast of NSW, Cornish & Binns (1987) and Mackay & Robinson (1987) observed changes in stream water quality following a wildfire that burnt part of a replicated paired catchment study. In the initial post-fire period, stream water cationic concentrations decreased relative to an unburnt control catchment and this impact was attributed to a relative increase in discharge in the burnt catchments. However, post-fire, cationic concentrations had increased relative to the pre-fire period after 3–4 years. Potassium concentrations increased in the year of the fire, as did Ca concentrations (Mackay & Robinson, 1987), yet turbidity levels remained unchanged. Due to heavy groundcover regeneration following the fire, turbidity levels in fact decreased to significantly below control and pre-fire levels five years after the catchments had been burnt (Cornish & Binns, 1987).

Following the 2001/2 Sydney fires, Wallbrink *et al.* (2004) reported that wildfires can potentially have a significant impact on downstream water quality within Lake Burragorang, the largest reservoir forming part of Sydney's water supply. The extensive wildfires in some areas were of extreme intensity, particularly on ridge-tops and flat–moderate (0–11°) northwesterly facing slopes, producing heat energy levels >70 000 kW m^{-1} (Chafer *et al.*, 2004). Varying effects on soil water repellency were observed ranging from no change to enhancement or destruction depending upon the soil temperatures attained (Shakesby *et al.*, 2003). In early 2002, a series of rainfall events post-fire resulted in accelerated soil erosion, redistribution of significant volumes of sediment on slopes and the storage and deposition of ash and sediment in rivers within the Nattai catchment (Blake *et al.*, 2009). High amplitude discharge peaks were observed, as were high turbidity, nitrogen and phosphorus concentrations (Wallbrink *et al.*, 2004; Blake *et al.*, 2009). Post-fire nitrogen, phosphorus and suspended sediment concentrations remained elevated for at least five years (Wilkinson *et al.*, 2006). However, post-fire bioturbation and litter dams on the foot slopes limited further downstream transport and propagation of sediment and nutrients (Shakesby *et al.*, 2007). Nonetheless, sediment deposition was evident in deltas where tributaries enter Lake Burragorang, but was lower in magnitude than the long-term pre-fire annual mean due to drought conditions and low annual water yields (Wilkinson *et al.*, 2006, 2007). It was concluded, however, that under worst-case scenario conditions, post-fire annual sediment yields to Lake Burragorang could be two to three orders of magnitude higher than the typical mean annual yield (Wilkinson *et al.*, 2007).

Despite some evidence that wildfires may not play a major geomorphic role in southeastern Australia (e.g. Tomkins *et al.*, 2007), Shakesby *et al.* (2007) stated that "*post-fire transfer of large proportions of the topsoil and its nutrients to watercourses has deleterious effects on downstream water quality*". In light of this and the destruction caused by the Black Christmas fires of 2001/2, a parliamentary inquiry has resulted in a greater focus on hazard reduction activities (NSW Parliament, 2002). This has been bolstered by more recent predictions of a 20–84% increase in potential large-fire ignition days under projected 2050 climate scenarios (Bradstock *et al.*, 2009). However, given that forests in NSW occur on a mixture of land tenures, legal, institutional and economic barriers limit the effectiveness of efforts to reduce wildfire risk with consequences for small and large water supplies.

Against the above context, this paper aims to outline current forest management within water supply catchments in NSW highlighting improvements required to reduce the impact of wildfires on river ecosystems, water supply and treatment. It further aims to introduce the concept of payments for ecosystem services (PES) by presenting two case studies from the USA where PES schemes are successfully being implemented to reduce the risk of wildfires in source water supply catchments. Feasible changes to existing management and institutional frameworks are evaluated for possible adoption in NSW.

MANAGEMENT OF FORESTS, FIRES AND WATER SUPPLY CATCHMENTS IN NSW

Water utilities

In NSW, there are 112 water utilities that supply water to cities and regional centres (Fig. 1). The bulk of these are local water utilities owned and operated by local councils delivering water to a combined regional population of 1.8 million (DPI, 2012). The two largest cities are served by the major state-owned water utilities, Sydney Water Corporation (Sydney) and Hunter Water Corporation (Newcastle), delivering water to populations of 4.4 million and 560 000, respectively. In addition to the state and local council operated water supplies, there are 19 independent water associations in NSW serving small regional communities (Webb, 2012a).

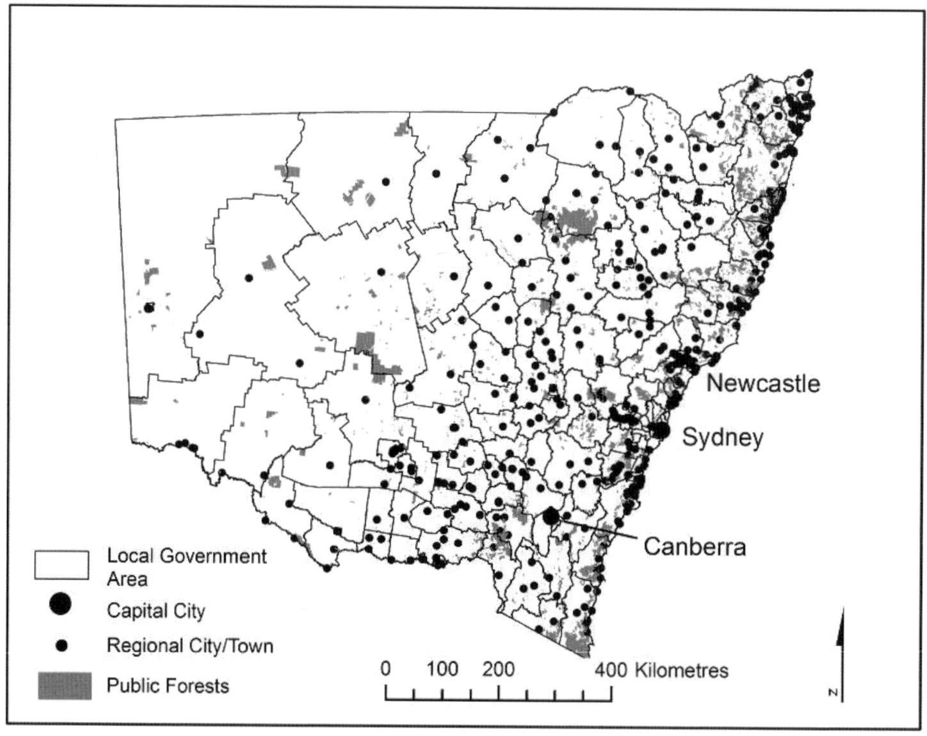

Fig. 1 The extent of publicly owned forests in relation to cities and local government areas in NSW. Note the extent of privately owned/leased forests is not shown.

Forest and catchment management

As mentioned in the Introduction, forests in NSW are located on a mixture of land tenures. In the majority of water supply catchments, the water utilities or catchment authorities rarely have control over the majority of the forests supplying water to streams and reservoirs (Webb, 2012a). This is even the case in the extensive 1.6 million ha catchments managed by the Sydney Catchment Authority (SCA), which delivers bulk raw water to the Sydney Water Corporation. The SCA has principal responsibility for managing the "special areas" closest to the reservoirs that mostly comprise native eucalypt forest. However, these areas comprise only 20% of the total catchments and the majority is held in conservation reserves managed subject to a memorandum of understanding between the SCA and the National Parks division of the NSW Office of Environment and Heritage (OEH) (SCA, 2007). The remainder of the catchments are a mixture of National Parks, State forests, Crown lands and private property over which the SCA has limited control. In the case of local council water utilities, the majority of their catchments comprise a mixture of private and state-owned lands over which they have minimal or zero jurisdiction.

The limited control of water utilities over forest and other land management within drinking water catchments poses problems for the effective management of water supplies in NSW. This is, arguably, exacerbated by the lack of an overall standard for the quality of water entering reservoirs and supply systems in Australia (McKay & Moeller, 2001). There is also limited scope for water authorities to promote or fund forest and other land management practices aimed at improving inflow water quality.

Fire management

To complicate matters further, fires are managed by four separate state government agencies. Fires on National Parks fall within the jurisdiction of the National Parks and Wildlife Division of OEH; Forests NSW has jurisdiction over fires on State forests and Crown timber-lands; the NSW Fire Brigade has control over metropolitan fires, while the NSW Rural Fire Service is tasked with providing services for the prevention, mitigation and suppression of fires in rural fire districts (Gillen, 2005).

The SCA has developed joint fire management plans for the special areas within the Sydney catchments with OEH (SCA, 2007). These plans cover activities such as fire trail maintenance, hazard reduction burning and rapid response strategies in the event of wildfires. However, as mentioned, SCA does not have control over the remaining 80% of the catchments and is reliant upon other landowners to manage fires. Hunter Water Corporation also acknowledges the risk that wildfires potentially pose for water quality and is in the process of developing a bushfire management plan in consultation with other stakeholders (Hunter Water, 2011). In the case of local water utilities, they do not have any jurisdiction over fire management and so potential effects on water quality are at the whim of hazard reduction and suppression activities undertaken by State government agencies. Furthermore, the majority of water utilities are small and lack the funds and resources (Woodbury & Dollery, 2004) to contribute to reducing the risk of wildfires.

PAYMENTS FOR ECOSYSTEM SERVICES: AN ALTERNATIVE APPROACH

"Ecosystem services" are goods and services provided by ecosystems for the benefit of humans (Daily, 1997; MEA, 2005). A specific category of ecosystem services that relates purely to catchments or watersheds has been termed "hydrologic services" (Brauman *et al.*, 2007) or "watershed services" (Stanton *et al.*, 2010). Examples include the provision of water for domestic, agricultural, commercial, industrial and hydro-electric power generation purposes; the supply of fish and other freshwater products; reduction of flood damage; prevention of soil erosion and sedimentation of waterways and reservoirs; and recreation and aesthetic values (Brauman *et al.*, 2007).

Broadly defined, payments for ecosystem services (PES) are a payment or exchange of credits between a buyer and a seller to effect some improvement of an ecosystem service (Stanton *et al.*, 2010). A subset of PES schemes dealing with watershed services is called payments for watershed services (PWS) schemes and these can be grouped into Government PWS, Private PWS and Water Quality Trading (WQT) schemes (Webb, 2012b). Government PWS schemes involve a government agency as the buyer of watershed services, sometimes on behalf of others, from upstream land owners. Payments take many forms including economic incentives to change land use practices, subsidies, cost-sharing arrangements, tax relief, land purchase deals and the purchase of conservation easements (Webb, 2012b). Private PWS schemes are rare and involve payments wholly by a private entity to upstream land owners to protect a watershed service for either business reasons or philanthropic interests. Water quality trading schemes involve a framework whereby a cap on pollution loads is set by regulation and the regulated entities purchase and trade in offset credits to meet their obligations (Shortle & Horan, 2001).

The number of PWS schemes, and the area of land protected, is expanding worldwide. In the USA alone in 2008, PWS transactions exceeded US$1.3 billion and operated across 16.4 million ha (Stanton *et al.*, 2010). While these schemes operate by exchanging payments or credits for watershed services in one way or another, there is often a driver leading to their existence and ultimate success or failure. Typically the drivers are regulatory and/or economic (Webb, 2012b). In the USA, an interesting development has been the introduction of PES/PWS schemes to reduce the risk of wildfires in water supply catchments. Two such case studies are presented below.

Santa Fe, New Mexico

The city of Sante Fe, New Mexico, has adopted an innovative scheme to reduce the risk of wildfires within its water supply catchment. The author visited the city and water supply catchment in April 2011 and conducted interviews with personnel from the US Forest Service (USFS) and Santa Fe Watershed Association.

The Santa Fe region experiences a cool semi-arid climate where wildfires are not uncommon (Steelman & Kunkel, 2004; Gutzler & Van Alst, 2010). The city lies at an elevation of 2130 m at the foot of the Sangre de Cristo Mountains and today has a population of ~75 000. The Santa Fe Municipal watershed supplies up to 50% of the water used, the remainder coming from groundwater wells. The watershed is ~6954 ha in area, of which the City owns and manages around 400 ha near the dams and riparian zones. The rest of the watershed comprises the Santa Fe National Forest in the lower watershed and the Pecos Wilderness Area (~4000 ha) in higher parts. These forested areas are managed by the USFS and not the City.

For more than 100 years, the Santa Fe watershed forests have been managed with a philosophy of complete fire suppression and this has led to the stands becoming overstocked, vulnerable to pest or insect attack, and highly prone to intense stand-replacement fires with drastic consequences for soil erosion and contamination of the local water supply (Santa Fe, 2009). Having witnessed the effects of the Hayman and Buffalo Creek fires on water supplies in Denver, Colorado (e.g. Moody & Martin, 2001; Robichaud *et al.*, 2008), as well as the 2000 Cerro Grande fire that resulted in a 140-fold increase in sedimentation of the Los Alamos reservoir (Reneau *et al.*, 2007), a consensus decision was made that Santa Fe needed to act. Discussions began in 2007 and culminated in the Santa Fe Municipal Watershed 20-Year Protection Plan which is unique in that it "*seeks to fund forest restoration activities using the payments for ecosystem services model as an insurance policy against future threats, particularly of catastrophic fire, to the municipal water supply*" (Santa Fe, 2009). The major signatories are the USFS, Santa Fe Watershed Association, the Nature Conservancy, City of Santa Fe Water Division and the City of Santa Fe Fire Department. The cost of the Plan is estimated to be US$4.3 million over 20 years. By comparison, investing in the catchment would avoid an estimated repair bill of US$22 million if 2800 ha of the watershed were to burn. Without hazard reduction, the likelihood of such a fire was estimated to be one in five in any given year (Santa Fe, 2009).

The four critical components of the Plan are vegetation management and fire use; water management; public awareness and outreach; and financial management based on PES. Between 2003 and 2006, mechanical treatments were carried out across 2114 ha of forests in the areas dominated by Ponderosa pines (<3050 m). The pines were thinned using a "mastication" method that reduces trees to large strips or chunks of wood that can more easily be burnt as part of hazard reduction. The aim of the thinning and mechanical treatments is to separate the crowns to reduce the risk of crown fires (Fig. 2). Under the Plan, some additional mechanical works have been outlined which include the targeted reduction of undergrowth to reduce the density of piñon-juniper woodland.

In the Santa Fe National Forest, there is a plan to undertake regular hazard reduction burning across the already treated areas. The aim is to burn approximately 400 ha per year with the entire area of the lower watershed being burned once every seven years. In the Pecos Wilderness Area, which covers predominantly the highest elevation areas in the watershed, Federal regulations preclude any mechanical thinning or similar interventions. However, prescribed burning can be undertaken. The wilderness area at lower elevations comprises mixed conifer forests, including Ponderosa pines, Piñon pines and Gambel oak. At these elevations, there are plans to undertake prescribed burns to reduce fuel loads. Under the regulations it is also possible to undertake strategic hand thinning to break up fuels and this will be considered. To somewhat protect the wilderness area, mechanical thinning will be undertaken on ridges adjacent to its borders. At higher elevations (>3050 m) in the wilderness area, the predominant forests are spruce and fir stands within which active management activities are not planned (Santa Fe, 2009).

The Plan is at first being funded by federal funds obtained from Congress allocations. However, from 2014 the USFS's work in the watershed will need to be funded by cost-sharing arrangements. To that end, the plan is to establish a PES scheme that leverages US$0.65 per month from water users in the City as part of their service rate charges. Underpinning much of the selling of the scheme is the public education and outreach programme. This is being handled by the Santa Fe Watershed Association and includes a number of initiatives including taking the general public on supervised educational hikes through the watershed, conducting field days with school groups, presentations to primary school classes, involving high school students in water monitoring activities, staffing educational tables in the city, distribution of newsletters, placing an information page in the telephone book and developing 30-second television advertisements (Santa Fe, 2009).

According to surveys conducted on behalf of the Nature Conservancy, the PES scheme is supported by 76% of Santa Fe residents (Felicity Broennan, written communication, 14 July 2011). Based on this level of support, the PES scheme is likely to be a success. However, despite the investment and work undertaken, there can be no guarantee that a wildfire will not occur. If it does, the "insurance" work is designed to mitigate its impact upon the water supply.

Fig. 2 Stands of Ponderosa Pine recently thinned to separate the crowns and reduce wildfire risk in the Santa Fe watershed.

Denver, Colorado

A brief summary of the PES scheme operating in Denver's water supply catchments is provided here as it has been described in greater detail by Webb (2012b). Denver Water is a non-profit public utility that supplies drinking water to around 1.3 million people. Its watersheds total one million ha across four river basins delivering water to 15 major reservoirs. Much of the land in the watersheds is forested public land managed by the USFS and not Denver Water.

The PES scheme has originated in response to a number of large wildfires experienced in the watersheds in recent decades (Moody & Martin, 2001; Robichaud *et al.*, 2008). The Hayman fire alone caused US$38 million in property damage, cost US$42 million to suppress, and between Denver Water and the USFS cost US$48 million to remediate. There was an acknowledgement from both agencies that management of National forest land was impacting in an adverse way on Denver Water's business. Denver Water had no jurisdiction over such land and could not ensure the USFS would undertake appropriate management. The USFS was aware of Denver Water's requirements, but could not undertake the required management due to limited appropriated funds.

Following much negotiation, on 29 July 2010 Denver Water and the USFS signed a five-year Memorandum of Understanding titled "Restoring forest and watershed health to protect the City and County of Denver's municipal water supplies and infrastructure". The deal will see Denver Water and the USFS each contribute US$16.5 million towards restoring various watersheds. Hazard reduction activities such as forest thinning and rehabilitation of previously burnt areas will occur. The PES scheme will be funded in part by payments of US$27 per household in total over five years by Denver Water's customers. Denver Water has been running a comprehensive education campaign and the USFS is committed to engaging communities in forest and conservation management projects. The combined effort of these agencies has contributed to substantial local support for the PES scheme (Webb, 2012b).

CONCLUSIONS

Recent history and scientific investigations indicate that extreme wildfires can and do occur in the forests of NSW, with potentially disastrous consequences for water supplies essential for a range of stakeholders. The existing management of forests, water supplies and fires within NSW is complex and there exist institutional, economic and legal barriers to the effective reduction of wildfire risk within source water supply catchments. Principal among these is the inherent lack of jurisdiction of water utilities over forest and fire management.

Payments for ecosystem services schemes offer an alternative approach, whereby downstream users of the water resource pay additional water rates that are used to fund appropriate hazard reduction activities to reduce the risk and severity of wildfires. Initiation of such schemes by water utilities in Santa Fe and Denver demonstrates that market instruments can be utilised to overcome the identified institutional, economic and legal barriers. Similar schemes have been successfully implemented around the world to reduce the impacts of land use (e.g. O'Grady, 2011) and wastewater treatment (e.g. Cochran & Logue, 2011), with water quality improvements often negating the need for expensive treatment plants (e.g. Salzman, 2011). Payments for ecosystem services scheme funds and appropriate memoranda of understanding could successfully be used by NSW water utilities, large and small, to pay forest and fire management agencies for hazard reduction activities to improve water quality.

Acknowledgements This work was funded by a 2011 Gottstein Fellowship with support from Forests NSW. The author is indebted to Sandy Hurlocker, Felicity Broennan, Claire Harper, Polly Hays, Tommy John, Joan Carlson, Susie Weingardt, Rick Cables and Marc Waage for generously hosting him and providing insights into the two case studies presented.

REFERENCES

ABARES (2011) *Australia's Forests at a Glance 2011*. Australian Bureau of Agricultural and Resource Economics and Sciences, Canberra.

Atkinson, G. (1984) Erosion damage following bushfires. *Journal of the Soil Conservation Service of NSW* 40, 4–9.

Blake, W. H., Wallbrink, P. J., Wilkinson, S. N., Humphreys, G. S., Doerr, S. H., Shakesby, R. A. & Tomkins, K. M. (2009) Deriving hillslope sediment budgets in wildfire-affected forests using fallout radionuclide tracers. *Geomorphology* 104, 105–116.

Blong, R. J., Riley, S. J. & Crozier, P. J. (1982) Sediment yield from runoff plots following bushfire near Narrabeen Lagoon, NSW. *Search* 13, 36–38.

Bradstock, R. A., Cohn, J. S., Gill, A. M., Bedward, M. & Lucas, C. (2009) Prediction of the probability of large fires in the Sydney region of south-eastern Australia using components of fire weather. *Int. J. Wildland Fire* 18, 932–943.

Brauman, K. A., Daily, G. C., Ka'eo Duarte, T. & Mooney, H. A. (2007) The nature and value of ecosystem services: an overview highlighting hydrologic services. *Ann. Rev. Environ. Res.* 32, 67–98.

Chafer, C. J., Noonan, M. & Macnaught, E. (2004) The post-fire measurement of fire severity and intensity in the Christmas 2001 Sydney wildfires. *Int. J. Wildland Fire* 13, 227–240.

Cheney, N. P. (1995) Bushfires – an integral part of Australia's environment. In: *Year Book Australia No. 77*, 515–521. Australian Bureau of Statistics, Canberra.

Cochran, B. & Logue, C. (2011) A watershed approach to improve water quality: case study of Clean Water Services' Tualatin River program. *J. Am. Water Resour. Assoc.* 47(1), 29–38.

Coghlan, B. J. (2004) The human health impact of the 2001–2002 "Black Christmas" bushfires in New South Wales, Australia, an alternative multidisciplinary strategy. *Journal of Rural and Remote Environmental Health* 3, 18–28.

Cornish, P. M. & Binns, D. (1987) Streamwater quality following logging and wildfire in a dry sclerophyll forest in southeastern Australia. *Forest Ecology and Management* 22, 1–28.

Daily, G. C. (1997) *Nature's Services: Societal Dependence on Natural Ecosystems*. Island Press, Washington DC.

DPI (2012) Local water utilities. NSW Office of Water, Department of Primary Industries, http://www.water.nsw.gov.au/Urban-water/Local-water-utilities/default.aspx (accessed 19 January 2012).

Dragovich, D. & Morris, R. (2002a) Fire intensity, runoff and sediment movement in eucalypt forest near Sydney, Australia. In: *Applied Geomorphology: Theory and Practice* (ed. by R. J. Allison), 145–164. Wiley, Chichester.

Dragovich, D. & Morris, R. (2002b) Fire intensity, slopewash and bio-transfer of sediment in eucalypt forest, Australia. *Earth Surf. Processes Landf.* 27, 1309–1319.

Gillen, M. (2005) Urban governance and vulnerability: exploring the tensions and contradictions in Sydney's response to bushfire threat. *Cities* 22, 55–64.

Gutzler, D. S. & Van Alst, L. (2010) Interannual variability of wildfires and summer precipitation in the Southwest. *New Mexico Geology* 32, 22–24.

Hunter Water (2011) *Catchment Management Plan: Hunter Water's Eight Element Plan for Our Catchments*. Hunter Water Corporation, Newcastle, Australia.

Mackay, S. M. & Robinson, G. (1987) Effects of wildfire and logging on streamwater chemistry and cation exports of small forested catchments in southeastern New South Wales, Australia. *Hydrol. Processes* 1, 359–384.

McKay, J. & Moeller, A. (2001) Is risk associated with drinking water in Australia of significant concern to justify mandatory regulation. *Environmental Management* 28, 469–481.

MEA (2005) *Ecosystems and Human Well-being: Synthesis*. Island Press, Washington DC.

Moody, J. A. & Martin, D. A. (2001) Initial hydrologic and geomorphic response following a wildfire in the Colorado Front Range. *Earth Surf. Processes Landf.* 26, 1049–1070.

NSW Parliament (2002) *Report on the Inquiry into the 2001/2002 Bushfires*. Joint Select Committee on Bushfires, Sydney.

O'Grady, D. (2011) Sociopolitical conditions for successful water quality trading in the South Nation River watershed, Ontario, Canada. *J. Am. Water Resour. Assoc.* 47(1), 39–51.

Prosser, I. P. & Williams, L. (1998) The effect of wildfire on runoff and erosion in native *Eucalyptus* forest. *Hydrol. Processes* 12, 251–265.

Reneau, S. L., Katzman, D., Kuyumjian, G. A., Lavine, A. & Malmon, D. V. (2007) Sediment delivery after a wildfire. *Geology* 35, 151–154.

Robichaud, P. R., Wagenbrenner, J. W., Brown, R. E., Wohlgemuth, P. M. & Beyers, J. L. (2008) Evaluating the effectiveness of contour-felled log erosion barriers as a post-fire runoff and erosion mitigation treatment in the western United States. *Int. J. Wildland Fire* 17, 255–273.

Salzman, J. (2011) What is the emperor wearing? The secret lives of ecosystem services. *Pace Environmental Law Review* 28(2), 591–613.

Santa Fe (2009) *Santa Fe Municipal Watershed Plan 2010–2029*. Santa Fe, New Mexico.

SCA (2007) *Special Areas Strategic Plan of Management*. Sydney Catchment Authority, Penrith.

Shakesby, R. A., Chafer, C. J., Doerr, S. H., Blake, W. H., Wallbrink, P., Humphreys, G. S. & Harrington, B. A. (2003) Fire severity, water repellency characteristics and hydrogeomorphological changes following the Christmas 2001 Sydney forest fires. *Australian Geographer* 34, 147–175.

Shakesby, R. A., Wallbrink, P. J., Doerr, S. H., English, P. M., Chafer, C. J., Humphreys, G. S., Blake, W. H. & Tomkins, K. M. (2007) Distinctiveness of wildfire effects on soil erosion in south-east Australian eucalypt forests assessed in a global context. *Forest Ecology and Management* 238, 347–364.

Shortle J. S. & Horan R.D. (2001) The economics of nonpoint pollution control. *Journal of Economic Surveys* 15, 255–289.

Singh, G., Kershaw, A. P. & Clark, R. (1981) Quaternary vegetation and fire history in Australia. In: *Fire and the Australian Biota* (ed. by A. M. Gill, R. H. Groves & I. R. Noble), 23–54. Australian Academy of Science, Canberra.

Stanton, T., Echavarria, M., Hamilton, K. & Ott, C. (2010) *State of Watershed Payments: An Emerging Marketplace*. Forest Trends, Ecosystem Marketplace, Washington DC.

Steelman, T. A. & Kunkel, G. F. (2004) Effective community responses to wildfire threats: lessons from New Mexico. *Society & Natural Resources* 17, 679–699.

Tomkins, K. M., Humphreys, G. S., Wilkinson, M. T., Fink, D., Hesse, P. P., Doerr, S. H., Shakesby, R. A., Wallbrink, P. J. & Blake, W.H. (2007) Contemporary *versus* long-term denudation along a passive plate margin: the role of extreme events. *Earth Surf. Processes Landf.* 32, 1013–1031.

Wallbrink, P., English, P., Chafer, C., Humphreys, G., Shakesby, R., Blake, W. & Doerr, S. (2004) Impacts on water quality by sediments and nutrients released during extreme bushfires, Report 1: A review of the literature pertaining to the effect of fire on erosion and erosion rates, with emphasis on the Nattai catchment, NSW, following the 2001 bushfires. CSIRO Land & Water, Canberra.

Webb, A. A. (2012a) Can timber and water resources be sustainably co-developed in south-eastern New South Wales, Australia? *Environ. Dev. Sustain.* 14, 233–252.

Webb, A. A. (2012b) Payments for watershed services and the role of experimental catchment studies. In: *Revisiting Experimental Catchment Studies in Forest Hydrology* (ed. by A. A. Webb *et al.*), 207–216. IAHS Publ. 353. IAHS Press, Wallingford, UK.

Wilkinson, S., Wallbrink, P., Blake, W., Doerr, S. & Shakesby, R. (2006) Impacts on water quality by sediments and nutrients released during extreme bushfires, Report 3: Post-fire sediment and nutrient redistribution to downstream waterbodies, Nattai National Park, NSW. CSIRO Land & Water, Canberra.

Wilkinson, S., Wallbrink, P., Hancock, G., Blake, W., Shakesby, R. & Farwig, V. (2007) Impacts on water quality by sediments and nutrients released during extreme bushfires, Report 4: Impacts on Lake Burragorang. CSIRO Land & Water, Canberra.

Woodbury, K. & Dollery, B. (2004) Efficiency measurement in Australian Local Government: The case of New South Wales Municipal Water Services. *Review of Policy Research* 21, 615–636.

Zierholz, C., Hairsine, P. & Booker, F. (1995) Runoff and soil erosion in bushland following the Sydney bushfires. *Australian Journal of Soil and Water Conservation* 8, 28–37.

Effects of wildfire on source-water quality and aquatic ecosystems, Colorado Front Range

JEFFREY H. WRITER[1,2], R. BLAINE McCLESKEY[1] & SHEILA F. MURPHY[1]

1 *US Geological Survey, 3215 Marine Street, Boulder, Colorado 80303, USA*
jwriter@usgs.gov

2 *University of Colorado, Dept of Civil, Environmental, and Architectural Engineering, 428 UCB, Boulder, Colorado 80309, USA*

Abstract Watershed erosion can dramatically increase after wildfire, but limited research has evaluated the corresponding influence on source-water quality. This study evaluated the effects of the Fourmile Canyon wildfire (Colorado Front Range, USA) on source-water quality and aquatic ecosystems using high-frequency sampling. Dissolved organic carbon (DOC) and nutrient loads in stream water were evaluated for a one-year period during different types of runoff events, including spring snowmelt, and both frontal and summer convective storms. DOC export from the burned watershed did not increase relative to the unburned watershed during spring snowmelt, but substantial increases in DOC export were observed during summer convective storms. Elevated nutrient export from the burned watershed was observed during spring snowmelt and summer convective storms, which increased the primary productivity of stream biofilms. Wildfire effects on source-water quality were shown to be substantial following high-intensity storms, with the potential to affect drinking-water treatment processes.

Key words wildfire; Colorado Front Range; source-water quality; dissolved organic carbon; sediment; stream biofilms

INTRODUCTION

Effects of wildfire on municipal water supplies include earlier snowmelt from burned watersheds, post-fire erosion and associated transport of sediment to reservoirs and water treatment plants, and changes in source-water chemistry that may affect drinking-water treatability (Westerling *et al.*, 2006; Emelko *et al.*, 2010; Smith *et al.*, 2011). Challenges in evaluating wildfire effects on source-water quality and aquatic ecosystems include the lack of a pre-fire baseline data set, the unpredictable nature of the event, the remoteness of the terrain, and the difficulty of mobilizing scientific teams immediately after the fire to study watershed responses. It has become increasingly recognized that sampling regimes of high temporal frequency are necessary to characterize hydrological events in a representative and meaningful manner (Kirchner *et al.*, 2004; Pellerin *et al.*, 2011). Forested watersheds in the Colorado Front Range, USA, have a high risk of wildfire and post-fire erosion (Colorado State Forest Service, 2008), although studies evaluating the changes to source-water quality in these montane watersheds are limited to a few studies whose conclusions were based on monthly grab sampling (Hall and Lombardozzi, 2008; Rhoades *et al.*, 2011).

Post-wildfire, streams have shown increases in turbidity, nutrients, organic carbon, major ions and trace metals (Williams & Melack, 1997; Gresswell, 1999; Bladon *et al.*, 2008). Stream ecosystems play an important role in geochemical cycling (Allan & Costillo, 2007) and therefore affect carbon and nutrient export from watersheds (Battin *et al.*, 2003). Although nutrient cycling in streams is primarily controlled by stream biofilms (Paerl & Pickney, 1996), limited information exists on how this ecological function is affected by wildfire. Increased nutrients can stimulate development of stream biofilms, which consist of photosynthetic periphyton and heterotrophic microbes that are held together in an extra-cellular polymeric substrate (Paerl & Pinckney, 1996). Additionally, stream biofilms sequester suspended and dissolved inorganic and organic matter from the water column (Romani *et al.*, 2004). Previous studies by Stone *et al.* (2011) showed biofilm stabilization of bed sediments post-wildfire may reduce sediment export from burned watersheds.

In September 2010, the Fourmile Canyon Fire, located in the wildland–urban interface west of Boulder, Colorado, USA, burned over 2500 ha, destroyed more than 160 homes (USDA Forest Service, 2011), and threatened the water supply of the communities of Pinebrook Hills and

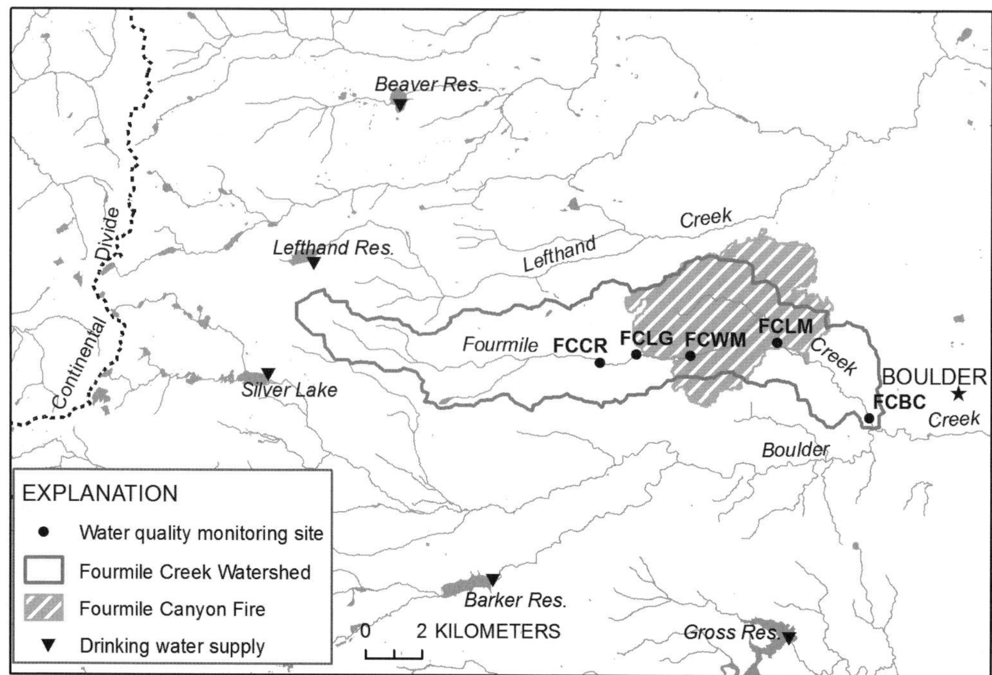

Fig. 1 Map of the Fourmile Creek watershed showing the extent of the 10–13 September 2010 wildfire and location of major water-supply reservoirs in the surrounding area.

Lafayette, which draw water from Fourmile and Boulder Creeks (Fig. 1). The primary focus of this study was to utilize high-frequency sampling to evaluate the effects of the Fourmile Canyon wildfire on water-quality and the stream ecosystem. Discharge in Fourmile Creek is dominated by spring snowmelt, although the summer convective storm season can contribute substantial short-term (less than 1 day) increases in flow.

METHODS AND MATERIALS

Precipitation, discharge and water-quality in Fourmile Creek were monitored for a year following the wildfire that occurred during 10–13 September 2010. Daily precipitation (P_{tot}) and 30-minute maximum rainfall intensity (normalized to an hour, I_{30}) at precipitation gauges, within and near the burned area, were used to characterize summer storm events (Murphy *et al.*, 2012). Water-quality monitoring sites were established at five locations on Fourmile Creek located upstream (FCCR, FCLG), within (FCWM, FCLM), and downstream (FCBC) of the burned area (Fig. 1). The watershed area and percentage of the watershed burned for each site are as follows: FCCR (26 km^2, 0%), FCLG (30 km^2, 0%), FCWM (38 km^2, 9%), FCLM (50 km^2, 25%), FCBC (63 km^2, 23%). Stream discharge was monitored at each site using a combination of methods that included discrete measurement and stream discharge estimates determined using stage measurements (for details see Murphy *et al.*, 2012). Water samples were analysed for nutrients, turbidity, major ions, trace metals, dissolved organic carbon (DOC), and UV-absorbance at 254 nm (UV_{254}) as described in McCleskey *et al.* (2012). Analytical variability based on replicate analyses of select samples for discussed parameters is as follows: nitrate (5%), DOC (2%) and UV_{254} (1%). Over specified time periods (e.g. during snowmelt runoff) statistical analysis of water samples was based on a comparison between samples collected at individual sites and on the same date using a paired t-test (Glantz, 2005).

To assess how observed increases in sediment loading, nutrients and DOC affected the stream ecosystem, the accumulation of stream biofilm on artificial substrates was monitored for three different periods post-wildfire (28 September–27 October 2010; 2 April–24 May 2011; and 21 September–19 October 2011) at sites FCCR, FCWM, and FCLM (Fig. 1) and at a reference site in Boulder Creek 2.5 km downstream from the confluence with Fourmile Creek. The Holm-Sidak multiple comparison test (Glantz, 2005) was used to statistically evaluate differences between stream biofilms at each site after 30 days.

RESULTS AND DISCUSSION

Stream discharge

Daily stream discharge (Fig. 2) was below the historical mean from April to mid-June as a result of limited snowfall in the watershed (Murphy *et al.*, 2012). In the spring of 2011, several frontal storm systems occurring in conjunction with the spring snowmelt resulted in a rapid rise in stream flow from a base flow of less than 0.06 $m^3 s^{-1}$ (2 cfs, cubic feet per second) up to a maximum measured value of 1.1 $m^3 s^{-1}$ (40 cfs). Following the spring snowmelt pulse, two substantial convective storms in July (daily precipitation 12–20 mm, maximum I_{30} 46 mm h^{-1}) produced short-term flash floods that resulted in stream discharge measurements increasing from less than 0.4 $m^3 s^{-1}$ (14 cfs) to greater than 23 $m^3 s^{-1}$ (800 cfs) during a span of less than 5 minutes (Murphy *et al.*, 2012).

Nutrient and organic carbon analysis of water samples

First flush After the wildfire, the first precipitation event occurred on 12 October 2010 (P_{tot} = 14 mm, I_{30} = 4 mm h^{-1}) and DOC concentrations substantially increased above base-flow levels (from 1.5 to 17 mg L^{-1} Fig. 2). Nitrate concentrations also increased above base-flow levels (<0.02 to 1.3 mg L^{-1}). Discharge was not measured during this storm.

Spring snowmelt DOC concentrations increased from 1 to 5 mg L^{-1} during spring snowmelt, but there was no significant difference ($p > 0.05$) between burned and unburned monitoring locations. Specific UV_{254} absorbance (SUVA, UV_{254}/dissolved organic carbon) was slightly higher ($p < 0.05$) in samples collected upstream from the burned area. Nitrate concentrations significantly increased ($p < 0.05$) at monitoring locations within and downstream from the burned area, during both spring snowmelt and low-intensity precipitation events. Ammonium concentrations were monitored throughout the study, but concentrations were generally below the detection limit (0.04 mg L^{-1}, McCleskey *et al.*, 2012).

Summer convective storms Two high-intensity convective storm events ($I_{30} > 40$ mm h^{-1}, Murphy *et al.*, 2012) occurred in the Fourmile Creek watershed in July 2011. The convective storms resulted in the mobilization and delivery of substantial amounts of sediment from hillslopes into Fourmile Creek. In addition, dramatic increases in DOC and nitrate concentrations were observed (DOC > 70 mg L^{-1}, nitrate > 9 mg L^{-1}) at monitoring locations within and below the burned area. The increased loading of DOC and nitrate to the stream in burned areas was more than two orders of magnitude greater than the loading from unburned areas. During convective storms, SUVA values were higher downstream from the burned areas, most likely the result of increased sediment and associated organic matter loading. As Fourmile Creek returned to base-flow discharge levels (less than 0.3 $m^3 s^{-1}$), turbidity (Murphy *et al.*, 2012), DOC and nitrate remained elevated compared to pre-storm base-flow water-quality. Additionally, a low-intensity storm on 7 September 2011 (daily precipitation 20 mm, I_{30} 8 mm h^{-1}; Murphy *et al.*, 2012) caused increases in DOC and nitrate concentrations that were not observed in pre-July storms of greater storm size and intensity. It appears lower intensity storms are remobilizing sediments deposited in the stream channel as a result of the high-intensity July convective storms, resulting in increased downstream concentrations of DOC and nitrate. It is likely that sediment stored in the stream

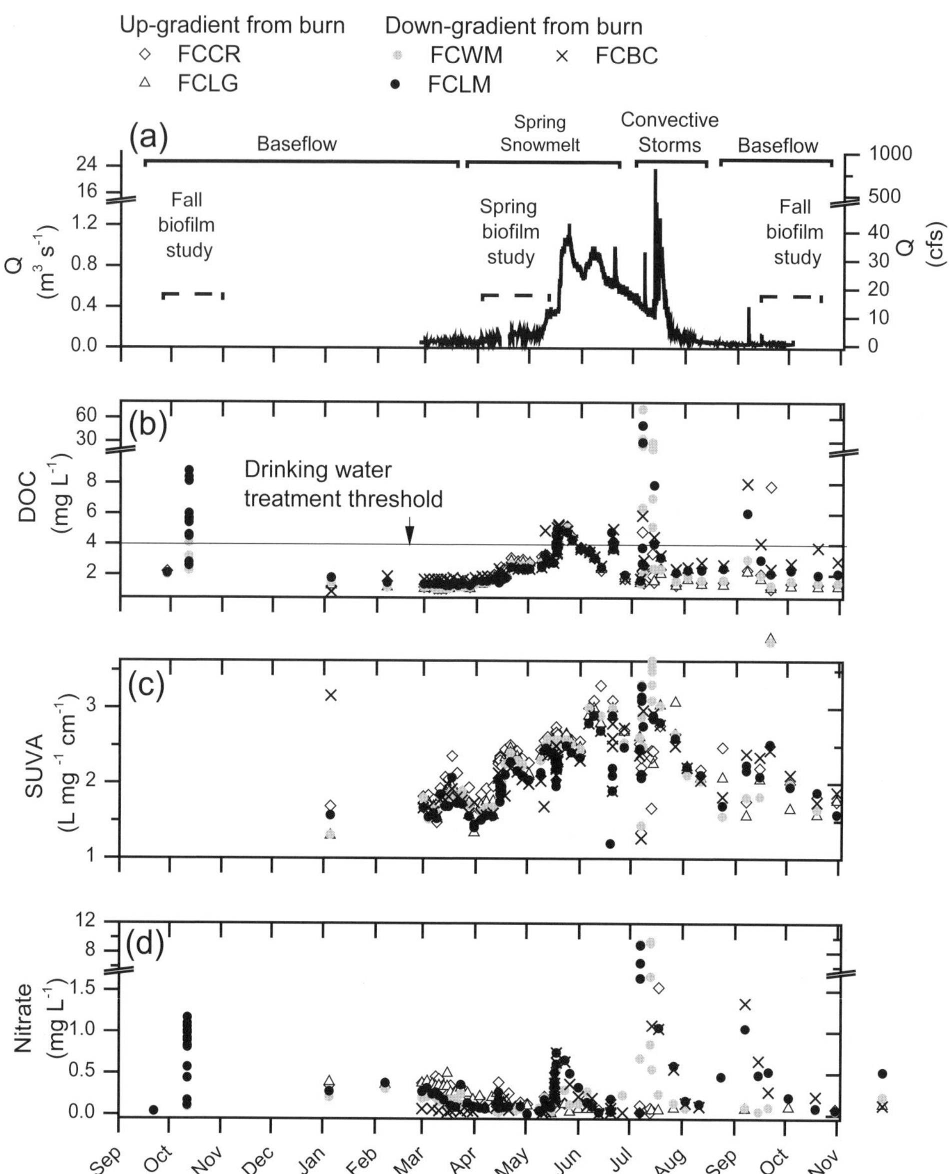

Fig. 2 (a) Stream discharge at FCLM, (b) DOC concentrations, (c) specific ultra-violet absorbance (SUVA) equivalent to UV absorbance at 254 nm normalized by DOC concentration, (d) nitrate concentrations.

channel will continue to affect water quality in the future. Research from other locales has shown wildfire-related effects on water-quality after four years of monitoring (Emelko *et al.*, 2011).

Effects of post-wildfire erosion on the stream ecosystem

The Autotrophic Index (*AI*, ratio between algal carbon and total organic matter) was used to evaluate changes in the stream biofilm community structure. After 30 days of exposure in the autumn of 2010 (after the wildfire, but before the convective storms), the Fourmile Creek sites had significantly (Holm-Sidak multiple comparison test, $p < 0.05$) lower *AI* values than the Boulder Creek reference site, indicative of nutrient limitations in Fourmile Creek. There were no significant differences in the *AI* values for all sites in the spring of 2011 exposure. However, in the autumn 2011 (post-convective storm) exposure, the *AI* values for the FCLM biofilms were similar to Boulder Creek biofilms, and significantly greater (Holm-Sidak multiple comparison test, $p < 0.05$) than the other Fourmile Creek biofilms. This increase may be indicative of increased nutrient supply benefitting the autotrophic community, and increases in both nitrate and sediment-associated phosphorus were observed (McCleskey *et al.*, 2012).

CONCLUSION AND IMPLICATIONS

This study focuses on discrete precipitation events following a wildfire and their corresponding effects on water-quality and the aquatic ecosystem. Increased erosion has been observed in several post-fire Colorado Front Range watersheds (e.g. Moody & Martin, 2001). Results from this study suggest that dramatic differences in sediment loading and associated water-quality can be expected between unburned and burned areas during high-intensity storm events. Stream discharge and turbidity increased by several orders of magnitude at monitoring locations downstream from the burned areas (Murphy *et al.*, 2012), meaning that the increased sediment-associated carbon and nutrient fluxes from burned watersheds have the potential to influence profoundly downstream aquatic ecosystems and watershed water supplies. The results also suggest that the primary productivity of the aquatic ecosystem increased as a result of increased sediment loading and associated nutrient supply (including both dissolved nitrogen species and sediment-bound phosphorus) and this response has been observed in other wildfire-affected watersheds (Emelko *et al.*, 2011; Smith *et al.*, 2011).

Acknowledgements The following individuals were invaluable to the completion of this investigation: George Aiken, Kenna Butler, Jennifer Carter, Brian Ebel, Bob Jarrett, Deborah Martin, John Moody, Deb Repert and Doug Winter (all with the USGS). Additionally, the following organizations were critical to the completion of this study; USGS Colorado Water Science Center, US Forest Service, City of Boulder, Pinebrook Hills Water District, Boulder County Parks & Open Space, Urban Drainage & Flood Control District, University of Colorado. This work was supported by the USGS National Research Program and the National Science Foundation (grant #0724960, the Boulder Creek Critical Zone Observatory). Any use of trade, firm, or product names is for descriptive purposes only and does not imply endorsement by the US Government.

REFERENCES

Allan, J. D. & Castillo, M. M. (2007) *Stream Ecology: Structure and Function of Running Waters*. Springer, Dordrecht, The Netherlands. 436 pp.

Battin, T. J., Kaplan, L. A., Newbold, J. D. & Hansen, C. M. E. (2003) Contributions of microbial biofilms to ecosystem processes in stream mesocosms. *Nature* 426, 439–442.

Bladon, K. D., Silins, U., Wagner, M. J., Stone, M., Emelko, M. B., Devito, K. J., Mendoza, C. A. & Boon, S. (2008). Wildfire impacts on nitrogen export and production from headwater streams in southern Alberta's Rocky Mountains. *Can. J. For. Res*. 38, 2359–2371.

Colorado State Forest Service (2008) Colorado Statewide Forest Resource Assessment – A Foundation for Strategic Discussion and Implementation of Forest Management in Colorado. http://csfs.colostate.edu/pages/statewide-forest-assessment.html. Accessed December 2011.

Emelko, M. B., Silins, U., Bladon, K. D. & Stone, M. (2011) Implications of land disturbance on drinking-water treatability in a changing climate: demonstrating the need for "source water supply and protection" strategies. *Water Res*. 45, 461–472.

Glantz, S. A. (2005) *Primer of Biostatistics*, 6th edn. McGraw-Hill, New York. 520 pp.

Gresswell, R. E. (1999) Fire and aquatic ecosystems in forested biomes of North America. *Trans. Am. Fisher. Soc.* 128, 193–221.

Hall, S. J. & Lombardozzi, D. (2008) Short-term effects of wildfire on montane stream ecosystems in the southern Rocky Mountains: One and two years post-burn. *Western North American Naturalist* 68, 453–462.

Kirchner, J. W., Feng, X., Neal, C. & Robson, A. J. (2004) The fine structure of water-quality dynamics: the (high-frequency) wave of the future. *Hydrol. Processes* 18, 1353–1359.

McCleskey, R. B., Writer, J. H. & Murphy, S. F. (2012) Water chemistry data for surface waters impacted by the Fourmile Canyon wildfire, Colorado, 2010–2011. *US Geological Survey Open-File Report* (in press).

Moody, J. A. & Martin, D. A. (2001) Initial hydrologic and geomorphic response following a wildfire in the Colorado Front Range. *Earth Surf. Processes Landf.* 26, 1049–1070.

Murphy, S. F., McCleskey, R. B. & Writer, J. W. (2012) Effects of flow regimes on stream turbidity and suspended solids after wildfire, Colorado Front Range, USA. In: *Wildfire and Water Quality: Processes, Impacts and Challenges* (Proc. conference held in Banff, Canada, June 2012). IAHS Publ. 354. IAHS Press. Wallingford, UK (this volume).

Paerl, H. W. & Pinckney, J. L. (1996) A mini-review of microbial consortia: Their roles in aquatic production and biogeochemical cycling. *Microb. Ecol.* 31, 225–247.

Pellerin, B. A., Saraceno, J. F., Shanley, J. B., Sebestyn, S. D., Aiken, G. A., Wollheim, W. M. & Bergamaschi, B. A. (2011) Taking the pulse of snowmelt: *In situ* sensors reveal seasonal, event and diurnal patterns of nitrate and dissolved organic matter variability in an upland forest stream. *Biogeochem.* doi:10.1007/s10533-011-9589-8.

Rhoades, C. C., Entwistle, D. & Butler, D. (2011) The influence of wildfire extent and severity on streamwater chemistry, sediment and temperature following the Hayman Fire, Colorado: *Int. J. Wildland Fire* 20, 430–442.

Romani, A. M., Guasch, H., Munoz, I., Ruana, J., Vilalta, E., Schwartz, T., Emtiazi, F. & Sabater, S. (2004) Biofilm structure and function and possible implications for riverine DOC dynamics. *Microb. Ecol.* 47, 316–328.

Smith, H. G., Sheridan, G. J., Lane, P. N. J., Nyman, P. & Haydon, S. (2011) Wildfire effects on water quality in forest catchments: a review with implications for water supply. *J. Hydrol.* 396, 170–192.

Stone, M., Emelko, M. B., Droppo, I. G. & Silins, U. (2011) Biostabilization and erodibility of cohesive sediment deposits in wildfire-affected streams. *Water Res.* 45, 521–534.

USDA Forest Service (2011) Fourmile Canyon Fire: Preliminary Findings. http://www.fs.fed.us/rmrs/docs/fourmile-canyon-fire/preliminary-findings.pdf. Accessed 5 March 2012.

Westerling A. L., Hidalgo, H. G., Cayan, D. R. & Swetnam, T. W. (2006) Warming and earlier spring increase Western U.S. forest wildfire activity. *Science* 313, 940–943.

Williams M. R. & Melack, J. M. (1997) Effects of prescribed burning and drought on the solute chemistry of mixed-conifer forest streams in the Sierra Nevada, California. *Biogeochem.* 39, 225–253.

Key word index